**HOLT**®

# ChemFile®

# Problem-Solving Workbook

**HOLT, RINEHART AND WINSTON**

A Harcourt Education Company

Orlando • **Austin** • New York • San Diego • Toronto • London

ISBN 0-03-036804-9

13  170  10

# Contents

# Basic Problem Solving for Chemistry

Chemistry is the science of matter—its properties and its changes. It is important that you learn how matter behaves and understand the reasons behind this behavior. Chemistry is also a quantitative science, which is what this book is all about. It is important that you learn the hows and whys of matter as well as the methods to determine or predict how much, how many, how concentrated, how big, or how hot—all of which involve measurement and calculation.

Here are some suggestions based on the firsthand experience of others that will help make you a successful problem solver in chemistry.

1. Make sure that you understand the ideas behind the problem. Few activities are more frustrating than trying to solve problems that deal with concepts you do not understand. To prevent this frustration, ask questions. Ask your teacher, ask other students, take advantage of any extra help available, and use any practice material available in your school's computer lab. Even better, ask the questions while your class is studying the chemistry concepts and before you are asked to solve problems.

2. Chemistry is a course that builds. You must master the basic ideas first because the next set of ideas will depend on your understanding of the previous set. In other words, learning new concepts in chemistry will increasingly depend on your mastery of what came before. Therefore, you should work very hard to gain understanding, especially in the early part of the course. If you do this, you will find that the second half of your chemistry course will be a smooth experience.

3. The problems in this book are solved by a four-step process that is explained in *The Mole and Chemical Composition*, although you will begin to use it in *The Science of Chemistry*. You should follow that process at least until you are confident that you can be successful with a system of your own.

4. Calculations in chemistry and other sciences use measured quantities. Therefore, the calculations are not just a mathematical exercise. The situations described by the problems will make more sense if you have an understanding of the units involved. How big is a liter? How much mass is represented by a gram? How hot is a Celsius temperature? If you can visualize these measurements, problems involving them will not seem so abstract.

5. Remember that your calculator is only a tool; it is not a problem solver. The answer that you get is only as valid as your method of solving the problem. Be sure to set up a problem carefully on paper before you start to punch in numbers on your calculator.

6. The best way to become good at solving chemistry problems is to practice. As in any other subject, the more you practice, the easier the problems will be. This book will give you a great deal of practice with the 24 topics covered in it. This combined with the problems and examples in your textbook should be enough to make you comfortable with solving basic chemistry problems.

**7.** The most important thing you can do to become a better problem-solver is to learn to organize your thoughts. This book will teach you to make tables of known and unknown information and to plan out solutions before putting numbers into your calculator. These methods of organization will help you in any subject in which you must solve problems.

**8.** Do not memorize the method of solution for problems. An exam may contain a problem that is similar to a problem in your textbook, but the test problem may be just different enough that your memorized way of solving will not work. It is far better to understand how to figure out a solution than to memorize how to solve each particular type of problem.

**9.** Do not expect to do well on a chapter test if you wait until the night before to work through the chapter. You must keep up with the material and study frequently, not just cram before an exam. Cramming is not an efficient method for learning, especially when you are trying to learn problem-solving skills.

**10.** Stay Calm! Chemistry can seem much more difficult when you are under stress. If you calmly approach chemistry problem-solving, you will have a much better chance of success.

Skills Worksheet

# Problem Solving

## Conversions

One of the aims of chemistry is to describe changes—to tell what changed, how it changed, and what it changed into. Another aim of chemistry is to look at matter and its changes and to ask questions such as how much, how big, how hot, how many, how hard, and how long did it take.

For example, chemistry asks the following:

- How much energy is needed to start a reaction?
- How much will the volume of a gas increase if you heat it?
- How long will a reaction take?
- How much can a reaction produce?
- How much of the reactant is needed to produce a required amount of product?
- How much energy does a reaction release?
- How high will the temperature of the solution get as a reaction occurs?

To answer these questions, chemists must make measurements. Measurements in science can never be treated as just numbers; they must always involve both a number and a *unit*. When you use measurements to calculate any quantity, the unit must always accompany the number in the calculation. Sometimes the unit given is not the most appropriate unit for the situation or calculation. In this case, a conversion can change the impractical unit into a more useful one. For instance, you would not want to measure the distance from New York City to San Francisco in inches. A simple conversion can transform the number of inches between the two cities to the much more practical number of miles.

**General Plan for Converting Measurements**

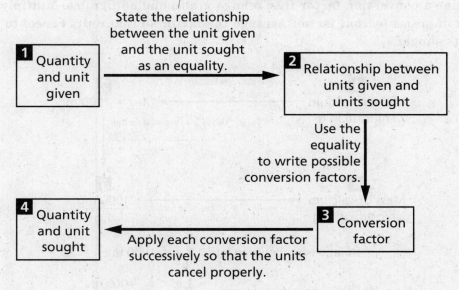

| Problem Solving *continued*

## CONVERTING SIMPLE SI UNITS

# Sample Problem 1

A small bottle contains 45.5 g of calcium chloride. What is the mass of calcium chloride in milligrams?

# Solution

### ANALYZE

What is given in the problem?    **the mass of calcium chloride in grams**

What are you asked to find?    **the mass of calcium chloride in milligrams**

A table showing what you know and what you do not know can help you organize the data. Being organized is a key to developing good problem solving skills.

| Items | Data |
|---|---|
| Quantity given | 45.5 g calcium chloride |
| Units of quantity given | grams |
| Units of quantity sought | milligrams |
| Relationship between units | 1 g = 1000 mg |
| Conversion factor | ? |
| Quantity sought | ? mg calcium chloride |

### PLAN

What steps are needed to convert grams to milligrams?

**Determine a conversion factor that relates grams and milligrams. Multiply the number of grams by that factor. Arrange the factor so that units cancel to give the units sought.**

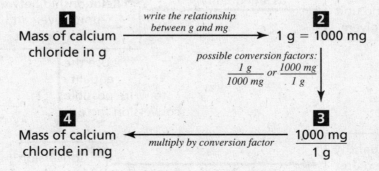

Relationship between units: 1 g = 1000 mg

Possible conversion factors: $\dfrac{\cancel{1\ g}}{\cancel{1000\ mg}}$ or $\dfrac{1000\ mg}{1\ g}$

**| Problem Solving** *continued*

The correct conversion factor is the one that when multiplied by the given quantity causes the units to cancel.

$$\underset{given}{\text{g calcium chloride}} \times \underset{1\ g}{\overset{\overset{\text{conversion factor}}{\overset{g \to mg}{1000\ mg}}}{}} = \underset{quantity\ sought}{\text{mg calcium chloride}}$$

**COMPUTE**

$45.5\ \cancel{g}\ \text{calcium chloride} \times \dfrac{1000\,mg}{1\ \cancel{g}} = 45\ 500\ \text{mg calcium chloride}$

**EVALUATE**

Are the units correct?

**Yes; milligrams are the desired units. Grams cancel to give milligrams.**

Is the answer reasonable?

**Yes; the number of milligrams is 1000 times the number of grams.**

# Practice

1. State the following measured quantities in the units indicated:

   **a.** 5.2 cm of magnesium ribbon in millimeters **ans: 52 mm**

   **b.** 0.049 kg of sulfur in grams **ans: 49 g**

   **c.** 1.60 mL of ethanol in microliters **ans: 1600 μL**

   **d.** 0.0025 g of vitamin A in micrograms **ans: 2500 μg**

   **e.** 0.020 kg of tin in milligrams **ans: 20 000 mg**

   **f.** 3 kL of saline solution in liters **ans: 3000 L**

**2.** State the following measured quantities in the units indicated:

  **a.** 150 mg of aspirin in grams **ans: 0.15 g**

  **b.** 2500 mL of hydrochloric acid in liters **ans: 2.5 L**

  **c.** 0.5 g of sodium in kilograms **ans: 0.0005 kg**

  **d.** 55 L of carbon dioxide gas in kiloliters **ans: 0.055 kL**

  **e.** 35 mm in centimeters **ans: 3.5 cm**

  **h.** 8740 m in kilometers **ans: 8.74 km**

  **i.** 209 nm in millimeters **ans: 0.000 209 mm**

  **j.** 500 000 μg in kilograms **ans: 0.0005 khhtg**

**3.** The greatest distance between Earth and the sun during Earth's revolution is 152 million kilometers. What is this distance in megameters? **ans: 152 000 Mm**

**| Problem Solving** *continued*

# Sample Problem 2

**A metallurgist is going to make an experimental alloy that requires adding 325 g of bismuth to 2.500 kg of molten lead. What is the total mass of the mixture in kilograms?**

# Solution

## ANALYZE

What is given in the problem?          **the mass of bismuth in grams, the mass of lead in kg**

What are you asked to find?          **the total mass of the mixture**

| Items | Data |
|---|---|
| Quantity given | 325 g of bismuth |
| Units of quantity given | grams |
| Units of quantity sought | kilograms |
| Relationship between units | 1000 g = 1 kg |
| Conversion factor | ? |
| Quantity sought | ? kg of bismuth |
| Mass of lead | 2.500 kg |
| Total mass | ? kg of mixture |

## PLAN

**To be added, the quantities must be expressed in the same units—in this case, kilograms. Therefore, 325 g of bismuth must be converted to kilograms of bismuth.**

What steps are needed to convert grams to kilograms?
**Determine a conversion factor that relates grams to kilograms. Apply that conversion factor to obtain the quantity sought.**

What steps are needed to find the total mass of the mixture in kilograms?
**Add the mass of the lead in kilograms to the mass of the bismuth in kilograms.**

## Problem Solving continued

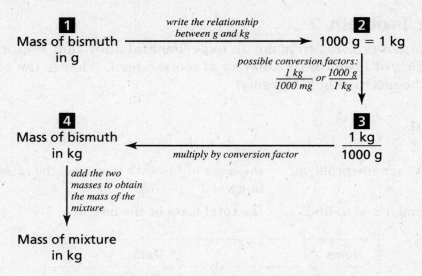

Relationship between units: 1000 g = 1 kg

Possible conversion factors: $\dfrac{\cancel{1000\ g}}{\cancel{1\ kg}}$ or $\dfrac{1\ kg}{1000\ g}$

$$\overset{given}{g\ bismuth} \times \overset{\overset{conversion\ factor}{g \rightarrow kg}}{\dfrac{1\ kg}{1000\ g}} = \overset{quantity\ sought}{kg\ bismuth}$$

$$\overset{calculated\ above}{kg\ bismuth} + \overset{given}{kg\ lead} = kg\ mixture$$

### COMPUTE

$$325\ \cancel{g}\ bismuth \times \dfrac{1\ kg}{1000\ \cancel{g}} = 0.325\ kg\ bismuth$$

$$0.325\ kg\ bismuth + 2.500\ kg\ lead = 2.825\ kg\ mixture$$

### EVALUATE

Are the units correct?
**Yes; kilograms are the units sought.**

Is the answer reasonable?
**Yes; the value, 0.325, is one-thousandth the given value, 325.**

**| Problem Solving** *continued*

# Practice

**1.** How many milliliters of water will it take to fill a 2 L bottle that already contains 1.87 L of water? **ans: 130 mL**

**2.** A piece of copper wire is 150 cm long. How long is the wire in millimeters? How many 50 mm segments of wire can be cut from the length? **ans: 1500 mm; 30 pieces**

**3.** The ladle at an iron foundry can hold 8500 kg of molten iron. 646 metric tons of iron are needed to make rails. How many ladlefuls of iron will it take to make 646 metric tons of iron? (1 metric ton = 1000 kg) **ans: 76 ladlefuls**

## CONVERTING DERIVED SI UNITS

## Sample Problem 3

A balloon contains 0.5 m$^3$ of neon gas. What is the volume of gas in cubic centimeters?

## Solution

### ANALYZE

What is given in the problem?    **the volume of neon in cubic meters**

What are you asked to find?      **volume of neon in cubic centimeters**

| Items | Data |
|---|---|
| Quantity given | 0.5 m$^3$ of neon |
| Units of quantity given | cubic meters |
| Units of quantity sought | cubic centimeters |
| Relationship between units | 1 m = 100 cm |
| Conversion factor | ? |
| Quantity sought | ? cm$^3$ neon |

### PLAN

What steps are needed to convert cubic meters to cubic centimeters?
**Rewrite the quantity in simple SI units. Determine the relationship between meters and centimeters. Write a conversion factor for each of the units in the given quantity. Multiply that quantity by the conversion factors. Arrange the factors so that units will cancel to give the units of the quantity sought.**

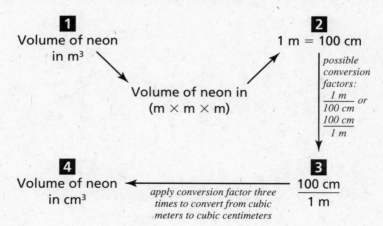

## | Problem Solving *continued*

Rewrite the given quantity in simple units as follows.

$$m^3 = m \times m \times m$$

Relationship between units: 1 m = 100 cm

Possible conversion factors: $\dfrac{1 \text{ m}}{100 \text{ cm}}$ or $\dfrac{100 \text{ cm}}{1 \text{ m}}$

*conversion factors cm → m,*
*applied three times*

$$\overset{given}{m^3 \text{ neon}} \times \frac{100 \text{ cm}}{1 \text{ m}} \times \frac{100 \text{ cm}}{1 \text{ m}} \times \frac{100 \text{ cm}}{1 \text{ m}} = \overset{quantity\ sought}{cm^3 \text{ neon}}$$

### COMPUTE

$$0.50 \text{ m}^3 \text{ neon} = 0.50 \,(m \times m \times m) \times \frac{100 \text{ cm}}{1 \text{ m}} \times \frac{100 \text{ cm}}{1 \text{ m}}$$
$$\times \frac{100 \text{ cm}}{1 \text{ m}} = 500\ 000 \text{ cm}^3 \text{ neon}$$

### EVALUATE

Are the units correct?

**Yes; cubic centimeters were the units sought.**

Is the answer reasonable?

**Yes; 500 000 cm$^3$ is half the number of cm$^3$ in 1 m$^3$.**

## Practice

**1.** State the following measured quantities in the units indicated.

**a.** 310 000 cm$^3$ of concrete in cubic meters **ans: 0.31 m$^3$**

**b.** 6.5 m$^2$ of steel sheet in square centimeters **ans: 65 000 cm$^2$**

**c.** 0.035 m$^3$ of chlorine gas in cubic centimeters **ans: 35 000 cm$^3$**

**d.** 0.49 cm$^2$ of copper in square millimeters **ans: 49 mm$^2$**

**| Problem Solving** *continued*

    **e.** 1200 dm$^3$ of acetic acid solution in cubic meters **ans: 1.2 m$^3$**

    **f.** 87.5 mm$^3$ of actinium in cubic centimeters **ans: 0.0875 cm$^3$**

    **g.** 250 000 cm$^2$ of polyethylene sheet in square meters **ans: 25 m$^2$**

**2.** How many palisade cells from plant leaves would fit in a volume of 1.0 cm$^3$ of cells if the average volume of a palisade cell is 0.0147 mm$^3$? **ans: 68 027 cells**

**▌Problem Solving** *continued*

# Additional Problems

**1.** Convert each of the following quantities to the required unit.

   **a.** 12.75 Mm to kilometers

   **b.** 277 cm to meters

   **c.** 30 560 $m^2$ to hectares (1 ha = 10 000 $m^2$)

   **d.** 81.9 $cm^2$ to square meters

   **e.** 300 000 km to megameters

**2.** Convert each of the following quantities to the required unit.

   **a.** 0.62 km to meters

   **b.** 3857 g to milligrams

   **c.** 0.0036 mL to microliters

   **d.** 0.342 metric tons to kilograms (1 metric ton = 1000 kg)

   **e.** 68.71 kL to liters

**3.** Convert each of the following quantities to the required unit:

   **a.** 856 mg to kilograms

   **b.** 1 210 000 μg to kilograms

   **c.** 6598 μL to cubic centimeters (1 mL = 1 $cm^3$)

   **d.** 80 600 nm to millimeters

   **e.** 10.74 $cm^3$ to liters

**4.** Convert each of the following quantities to the required unit:

   **a.** 7.93 L to cubic centimeters

   **b.** 0.0059 km to centimeters

   **c.** 4.19 L to cubic decimeters

   **d.** 7.48 $m^2$ to square centimeters

   **e.** 0.197 $m^3$ to liters

**5.** An automobile uses 0.05 mL of oil for each kilometer it is driven. How much oil in liters is consumed if the automobile is driven 20 000 km?

**6.** How many microliters are there in a volume of 370 $mm^3$ of cobra venom?

**7.** A baker uses 1.5 tsp of vanilla extract in each cake. How much vanilla extract in liters should the baker order to make 800 cakes? (1 tsp = 5 mL)

**8.** A person drinks eight glasses of water each day, and each glass contains 300 mL. How many liters of water will that person consume in a year? What is the mass of this volume of water in kilograms? (Assume one year has 365 days and the density of water is 1.00 kg/L.)

| **Problem Solving** *continued*

**9.** At the equator Earth rotates with a velocity of about 465 m/s.

   **a.** What is this velocity in kilometers per hour?

   **b.** What is this velocity in kilometers per day?

**10.** A chemistry teacher needs to determine what quantity of sodium hydroxide to order. If each student will use 130 g and there are 60 students, how many kilograms of sodium hydroxide should the teacher order?

**11.** The teacher in item 10 also needs to order plastic tubing. If each of the 60 students needs 750 mm of tubing, what length of tubing in meters should the teacher order?

**12.** Convert the following to the required units.

   **a.** 550 μL/h to milliliters per day

   **b.** 9.00 metric tons/h to kilograms per minute

   **c.** 3.72 L/h to cubic centimeters per minute

   **d.** 6.12 km/h to meters per second

**13.** Express the following in the units indicated.

   **a.** 2.97 kg/L as grams per cubic centimeter

   **b.** 4128 $g/dm^2$ as kilograms per square centimeter

   **c.** 5.27 $g/cm^3$ as kilograms per cubic decimeter

   **d.** 6.91 $kg/m^3$ as milligrams per cubic millimeter

**14.** A gas has a density of 5.56 g/L.

   **a.** What volume in milliliters would 4.17 g of this gas occupy?

   **b.** What would be the mass in kilograms of 1 $m^3$ of this gas?

**15.** The average density of living matter on Earth's land areas is 0.10 $g/cm^2$. What mass of living matter in kilograms would occupy an area of 0.125 ha?

**16.** A textbook measures 250. mm long, 224 mm wide, and 50.0 mm thick. It has a mass of 2.94 kg.

   **a.** What is the volume of the book in cubic meters?

   **b.** What is the density of the book in grams per cubic centimeter?

   **c.** What is the area of one cover in square meters?

**17.** A glass dropper delivers liquid so that 25 drops equal 1.00 mL.

   **a.** What is the volume of one drop in milliliters?

   **b.** How many milliliters are in 37 drops?

   **c.** How many drops would be required to get 0.68 L?

**Problem Solving** *continued*

**18.** Express each of the following in kilograms and grams:

 **a.** 504 700 mg

 **b.** 9 200 000 μg

 **c.** 122 mg

 **d.** 7195 cg

**19.** Express each of the following in liters and milliliters:

 **a.** 582 cm$^3$

 **b.** 0.0025 m$^3$

 **c.** 1.18 dm$^3$

 **d.** 32 900 μL

**20.** Express each of the following in grams per liter and kilograms per cubic meter.

 **a.** 1.37 g/cm$^3$

 **b.** 0.692 kg/dm$^3$

 **c.** 5.2 kg/L

 **d.** 38 000 g/m$^3$

 **e.** 5.79 mg/mm$^3$

 **f.** 1.1 μg/mL

**21.** An industrial chemical reaction is run for 30.0 h and produces 648.0 kg of product. What is the average rate of product production in the stated units?

 **a.** grams per minute

 **b.** kilograms per day

 **c.** milligrams per millisecond

**22.** What is the speed of a car in meters per second when it is moving at 100. km/h?

**23.** A heater gives off energy as heat at a rate of 330 kJ/min. What is the rate of energy output in kilocalories per hour? (1 cal = 4.184 J)

**24.** The instructions on a package of fertilizer tell you to apply it at the rate of 62 g/m$^2$. How much fertilizer in kilograms would you need to apply to 1.0 ha? (1 ha = 10 000 m$^2$)

**25.** A water tank leaks water at the rate of 3.9 mL/h. If the tank is not repaired, what volume of water in liters will it leak in a year? Show your setup for solving this. Hint: Use one conversion factor to convert hours to days and another to convert days to years, and assume that one year has 365 days.

**26.** A nurse plans to give flu injections of 50 μL each from a bottle containing 2.0 mL of vaccine. How many doses are in the bottle?

**Skills Worksheet**

# Problem Solving

## Significant Figures

A lever balance used to weigh a truckload of stone may be accurate to the nearest 100 kg, giving a reading of 15 200 kg, for instance. The measurement should be written in such a way that a person looking at it will understand that it represents the mass of the truck to the nearest 100 kg, that is, that the mass is somewhere between 15 100 kg and 15 300 kg.

Some laboratory balances are sensitive to differences of 0.001 g. Suppose you use such a balance to weigh 0.206 g of aluminum foil. A person looking at your data table should be able to see that the measurement was made on a balance that measures mass to the nearest 0.001 g. You should not state the measurement from the laboratory balance as 0.2060 g instead of 0.206 g because the balance was not sensitive enough to measure 0.0001 g.

To convey the accuracy of measurements, all people working in science use significant figures. *A significant figure is a digit that represents an actual measurement.* The mass of the truck was stated as 15 200 kg. The 1, 5, and 2 are significant figures because the balance was able to measure ten-thousands, thousands, and hundreds of kilograms. The truck balance was not sensitive enough to measure tens of kilograms or single kilograms. Therefore, the two zeros are not significant and the measurement has three significant figures. The mass of the foil was correctly stated as 0.206 g. There are three decimal places in this measurement that are known with some certainty. Therefore, this measurement has three significant figures. Had the mass been stated as 0.2060 g, a fourth significant figure would have been incorrectly implied.

### Rules for Determining Significant Figures

**A.** All digits that are not zeros are significant.

| All are nonzero digits. | All are nonzero digits. |
|:---:|:---:|
| ↓ ↓ ↓ | ↓ ↓ ↓ ↓ |
| 3 2 5 mL of ethanol | 1.3 2 5 g of zinc |
| **The measurement has three significant figures.** | **The measurement has four significant figures.** |

**B.** Zeros may or may not be significant. To determine whether a zero is significant, use the following rules:

## ▌Problem Solving *continued*

**1.** Zeros appearing between nonzero digits are significant.

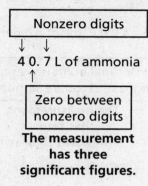

The measurement
has three
significant figures.

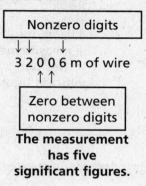

The measurement
has five
significant figures.

**2.** Zeros appearing in front of nonzero digits are not significant.

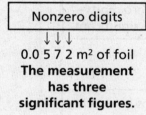

0.0 5 7 2 m² of foil
The measurement
has three
significant figures.

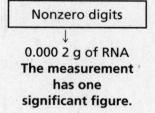

0.000 2 g of RNA
The measurement
has one
significant figure.

**3.** Zeros at the end of a number and to the right of a decimal are significant figures. Zeros between nonzero digits and significant zeros are also significant. This is a restatement of Rule 1.

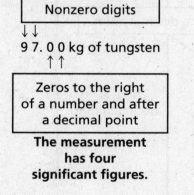

The measurement
has four
significant figures.

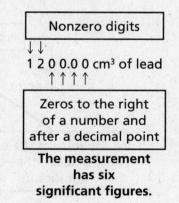

The measurement
has six
significant figures.

**4.** Zeros at the end of a number but to the left of a decimal may or may not be significant. If such a zero has been measured or is the first estimated digit, it is significant. On the other hand, if the zero has not been measured or estimated but is just a place holder, it is not significant. A decimal placed after the zeros indicates that they are significant.

## Problem Solving *continued*

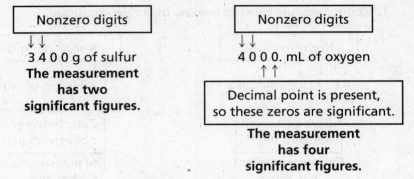

| Nonzero digits | Nonzero digits |
|---|---|
| ↓ ↓ | ↓ ↓ |
| 3 4 0 0 g of sulfur | 4 0 0 0. mL of oxygen |
| **The measurement has two significant figures.** | ↑ ↑ |
| | Decimal point is present, so these zeros are significant. |
| | **The measurement has four significant figures.** |

The rules are summarized in the following flowchart:

### General Plan for Determining Significant Figures

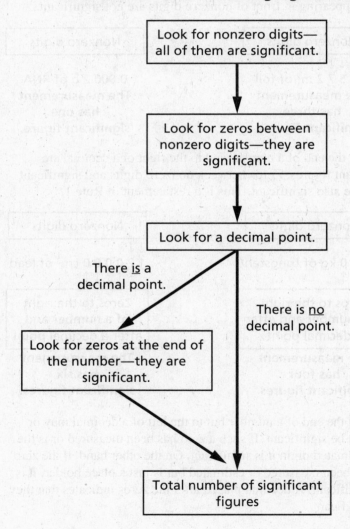

Look for nonzero digits— all of them are significant.

Look for zeros between nonzero digits—they are significant.

Look for a decimal point.

There **is** a decimal point.

There is **no** decimal point.

Look for zeros at the end of the number—they are significant.

Total number of significant figures

**| Problem Solving** *continued*

# Sample Problem 1

**Determine the number of significant figures in the following measure-ments:**

    **a. 30 040 g**
    **b. 0.663 kg**
    **c. 20.05 mL**
    **d. 1500. mg**
    **e. 0.0008 m**

# Solution
## ANALYZE

What is given in the problem?     **five measurements**

What are you asked to find?     **the number of significant figures in each meas-urement**

| Items | Data | | | | |
|---|---|---|---|---|---|
| | **a** | **b** | **c** | **d** | **e** |
| Measured quantity | 30 040 g | 0.663 kg | 20.05 L | 1500. mg | 0.0008 g |

## PLAN

What steps are needed to determine the number of significant figures in each measurement?
**Apply the steps in the flowchart to determine the number of significant figures.**

Apply the following steps from the flowchart. Eliminate the steps that are not applicable to the measurement in question.

| | |
|---|---|
| How many nonzero digits are there? | ? |
| How many zeros are there between nonzero digits? | ? |
| Is there a decimal point? | ? |
| How many significant zeros are at the end of the number? | ? |
| Total number of significant figures | ? |

## SOLVE

**a.** 30 040 g

| | |
|---|---|
| How many nonzero digits are there? | 2 |
| How many zeros are there between nonzero digits? | 2 |
| Is there a decimal point? | no |
| How many significant zeros are at the end of the number? | NA |
| Total number of significant figures | 4 |

The final zero is not significant.

**b.** 0.663 kg

| | |
|---|---|
| How many nonzero digits are there? | 3 |
| How many zeros are there between nonzero digits? | NA |
| Is there a decimal point? | yes |
| How many significant zeros are at the end of the number? | NA |
| Total number of significant figures | 3 |

The zero only locates the decimal point and is not significant.

**c.** 20.05 L

| | |
|---|---|
| How many nonzero digits are there? | 2 |
| How many zeros are there between nonzero digits? | 2 |
| Is there a decimal point? | yes |
| How many significant zeros are at the end of the number? | NA |
| Total number of significant figures | 4 |

**d.** 1500. mg

| | |
|---|---|
| How many nonzero digits are there? | 2 |
| How many zeros are there between nonzero digits? | NA |
| Is there a decimal point? | yes |
| How many significant zeros are at the end of the number? | 2 |
| Total number of significant figures | 4 |

There is a decimal following the final two zeros, so all digits are significant.

**Problem Solving** *continued*

**e.** 0.0008 g

| How many nonzero digits are there? | 1 |
|---|---|
| How many zeros are there between nonzero digits? | NA |
| Is there a decimal point? | yes |
| How many significant zeros are at the end of the number? | NA |
| Total number of significant figures | 1 |

The zeros are only place holders. They are not significant.

**EVALUATE**

Are the answers reasonable?

**Yes; all answers are in agreement with the rules for determining significant figures.**

# Practice

**1.** Determine the number of significant figures in the following measurements:

**a.** 640 cm$^3$ _____ans: 2_____

**b.** 200.0 mL _____ans: 4_____

**c.** 0.5200 g _____ans: 4_____

**d.** 1.005 kg _____ans: 4_____

**e.** 10 000 L _____ans: 1_____

**f.** 20.900 cm _____ans: 5_____

**g.** 0.000 000 56 g/L _____ans: 2_____

**h.** 0.040 02 kg/m$^3$ _____ans: 4_____

**i.** 790 001 cm$^2$ _____ans: 6_____

**j.** 665.000 kg·m/s$^2$ _____ans: 6_____

## DETERMINING SIGNIFICANT FIGURES IN CALCULATIONS

Suppose you want to determine the density of an ethanol-water solution. You first measure the volume in a graduated cylinder that is accurate to the nearest 0.1 mL. You then determine the mass of the solution on a balance that can measure mass to the nearest 0.001 g. You have read each measuring device as accurately as you can, and you record the following data:

| Measurement | Data |
|---|---|
| Mass of solution, $m$ | 11.079 g |
| Volume of solution, $V$ | 12.7 mL |
| Density of solution in g/mL, $D$ | ? |

## Problem Solving *continued*

You can determine density on your calculator and get the following result:

$$D = \frac{m}{V} = \frac{11.079 \text{ g}}{12.7 \text{ mL}} = 0.872\ 362\ 204 \text{ g/mL}$$

Although the numbers divide out to give the result shown, it is not correct to say that this quantity is the density of the solution. Remember that you are dealing with measurements, not just numbers. Consider the fact that you measured the mass of the solution with a balance that gave a reading with five significant figures: 11.079 g. In addition, you measured the volume of the solution with a graduated cylinder that was readable only to three significant figures: 12.7 mL. It seems odd to claim that you now know the density with an accuracy of nine significant figures.

You can calculate the density—or any measurement—*only as accurately as the least accurate measurement* that was used in the calculation. In this case the least accurate measurement was the volume because the measuring device you used was capable of giving you a measurement with only three significant figures. Therefore, you can state the density to only three significant figures.

### Rules for Calculating with Measured Quantities

| Operation | Rule |
|---|---|
| Multiplication and division | • Round off the calculated result to the same number of significant figures as the measurement having the fewest significant figures. |
| Addition and subtraction | • Round off the calculated result to the same number of decimal places as the measurement with the fewest decimal places. If there is no decimal point, round the result back to the digit that is in the same position as the leftmost uncertain digit in the quantities being added or subtracted. |

In the example given above, you must round off your calculator reading to a value that contains three significant figures. In this case, you would say:

$$D = \frac{m}{V} = \frac{11.079 \text{ g}}{12.7 \text{ mL}} = 0.872\ 362\ 204 \text{ g/mL} = 0.872 \text{ g/mL}$$

**▌Problem Solving** *continued*

# Sample Problem 2

**In an experiment to identify an unknown gas, it is found that 1.82 L of the gas has a mass of 5.430 g. What is the density of the gas in g/L?**

## Solution

### ANALYZE

What is given in the problem?　　　**the measured mass and volume of the gas**

What are you asked to find?　　　**the density of the gas**

| Items | Data |
|---|---|
| Mass of the gas, $m_{gas}$ | 5.430 g |
| Volume of the gas, $V_{gas}$ | 1.82 L |
| Density of the gas, $D_{gas}$ (numerical result) | ? g/L |
| Least number of significant figures in measurements | 3 (in 1.82 L) |
| Density of the gas, $D_{gas}$ (rounded) | ? g/L |

### PLAN

What step is needed to calculate the density of the gas?
**Divide the mass measurement by the volume measurement.**

What steps are necessary to round the calculated value to the correct number of significant figures?
**Determine which measurement has the fewest significant figures. Round the calculated result to that number of significant figures.**

$$D_{gas} = \frac{m_{gas}}{V_{gas}} = \text{numerical result} \xrightarrow{\substack{\text{round to correct} \\ \text{significant figures}}} \text{rounded result}$$

### COMPUTE

$$D_{gas} = \frac{m_{gas}}{V_{gas}} = \frac{\overset{\text{four significant figures}}{5.430 \text{ g}}}{\underset{\text{three significant figures}}{1.82 \text{ L}}} = \overset{\text{round to three significant figures}}{2.983\,\cancel{516\,484}} = 2.98 \text{ g/L}$$

*the digit following the 8 is less than 5, so the 8 remains unchanged*

### EVALUATE

Are the units correct?
**Yes; density is given in units of mass per unit volume.**

Are the significant figures correct?
**Yes; the mass had only three significant figures, so the answer was rounded to three significant figures.**

**| Problem Solving** *continued*

Is the answer reasonable?
**Yes; the mass/volume ratio is roughly 3/1, so the density is approximately 3 g/L.**

## Practice

1. Perform the following calculations, and express the result in the correct units and number of significant figures.

   **a.** $47.0 \div 2.2$ s **ans: 21 m/s**

   **b.** 140 cm $\times$ 35 cm **ans: 4900 cm$^2$**

   **c.** 5.88 kg $\div$ 200 m$^3$ **ans: 0.03 kg/m$^3$**

   **d.** 0.00 50 m$^2$ $\times$ 0.042 m **ans: 0.000 21 m$^3$**

   **e.** 300.3 L $\div$ 180. s **ans: 1.67 L/s**

   **f.** 33.00 cm$^2$ $\times$ 2.70 cm **ans: 89.1 cm$^3$**

   **g.** 35 000 kJ $\div$ 0.250 min **ans: 140 000 kJ/min**

## Sample Problem 3

Three students measure volumes of water with three different devices. They report the following volumes:

| Device | Volume measured |
|---|---|
| Large graduated cylinder | 164 mL |
| Small graduated cylinder | 39.7 mL |
| Calibrated buret | 18.16 mL |

If the students pour all of the water into a single container, what is the total volume of water in the container?

## Solution

### ANALYZE

What is given in the problem?   **three measured volumes of water**

What are you asked to find?   **the total volume of water**

| Items | Data |
|---|---|
| First volume of water | 164 mL |
| Second volume of water | 39.7 mL |
| Third volume of water | 18.16 mL |
| Total volume of water | ? |

### PLAN

What step is needed to calculate the total volume of the water?
**Add the separate volumes.**

What steps are necessary to round the calculated value to the correct number of significant figures?
**Determine which measurement has the fewest decimal places. Round the calculated result to that number of decimal places.**

### COMPUTE

$$V_{total} = V_1 + V_2 + V_3 = 164 \text{ mL} + 39.7 \text{ mL} + 18.16 \text{ mL}$$

$$
\begin{array}{r}
164 \text{ mL} \\
+ \ 39.7 \text{ mL} \\
+ \ 18.16 \text{ mL} \\
\hline
221.86 \text{ mL}
\end{array}
$$

Round the sum to the same number of decimal places as the measurement with the fewest decimal places (164 mL).

$$V_{total} = 221.86 \text{ mL} = 222 \text{ mL}$$

*the digit following the 1 is greater than 5, so the 1 is rounded up to 2*

**Problem Solving** *continued*

### EVALUATE

Are the units correct?

**Yes; the given values have units of mL.**

Are the significant figures correct?

**Yes; three significant figures is correct.**

Is the answer reasonable?

**Yes; estimating the values as 160, 40, and 20 gives a sum of 220, which is very near the answer.**

## Practice

1. Perform the following calculations and express the results in the correct units and number of significant figures:

   **a.** 22.0 m + 5.28 m + 15.5 m **ans: 42.8 m**

   **b.** 0.042 kg + 1.229 kg + 0.502 kg **ans: 1.773 kg**

   **c.** 170 cm$^2$ + 3.5 cm$^2$ − 28 cm$^2$ **ans: 150 cm$^2$**

   **d.** 0.003 L + 0.0048 L + 0.100 L **ans: 0.108 L**

   **e.** 24.50 dL + 4.30 dL + 10.2 dL **ans: 39.0 dL**

   **f.** 3200 mg + 325 mg − 688 mg **ans: 2800 mg**

   **g.** 14 000 kg + 8000 kg + 590 kg **ans: 23 000 kg**

**| Problem Solving** *continued*

# Additional Problems

**1.** Determine the number of significant figures in the following measurements:

  **a.** 0.0120 m

  **b.** 100.5 mL

  **c.** 101 g

  **d.** 350 cm$^2$

  **e.** 0.97 km

  **f.** 1000 kg

  **g.** 180. mm

  **h.** 0.4936 L

  **i.** 0.020 700 s

**2.** Round the following quantities to the specified number of significant figures:

  **a.** 5 487 129 m to three significant figures

  **b.** 0.013 479 265 mL to six significant figures

  **c.** 31 947.972 cm$^2$ to four significant figures

  **d.** 192.6739 m$^2$ to five significant figures

  **e.** 786.9164 cm to two significant figures

  **f.** 389 277 600 J to six significant figures

  **g.** 225 834.762 cm$^3$ to seven significant figures

**3.** Perform the following calculations, and express the answer in the correct units and number of significant figures.

  **a.** 651 cm × 75 cm

  **b.** 7.835 kg ÷ 2.5 L

  **c.** 14.75 L ÷ 1.20 s

  **d.** 360 cm × 51 cm × 9.07 cm

  **e.** 5.18 m × 0.77 m × 10.22 m

  **f.** 34.95 g ÷ 11.169 cm$^3$

**4.** Perform the following calculations, and express the answer in the correct units and number of significant figures.

  **a.** 7.945 J + 82.3 J − 0.02 J

  **b.** 0.0012 m − 0.000 45 m − 0.000 11 m

  **c.** 500 g + 432 g + 2 g

  **d.** 31.2 kPa + 0.0035 kPa − 0.147 kPa

  **e.** 312 dL − 31.2 dL − 3.12 dL

  **f.** 1701 kg + 50 kg + 43 kg

**5.** A rectangle measures 87.59 cm by 35.1 mm. Express its area with the proper number of significant figures in the specified unit:

  **a.** in cm$^2$

  **b.** in mm$^2$

  **c.** in m$^2$

## Problem Solving *continued*

6. A box measures 900. mm by 31.5 mm by 6.3 cm. State its volume with the proper number of significant figures in the specified unit:

   **a.** in cm$^3$

   **b.** in m$^3$

   **c.** in mm$^3$

7. A 125 mL sample of liquid has a mass of 0.16 kg. What is the density of the liquid in the following measurements?

   **a.** kg/m$^3$

   **b.** g/mL

   **c.** kg/dm$^3$

8. Perform the following calculations, and express the results in the correct units and with the proper number of significant figures.

   **a.** 13.75 mm × 10.1 mm × 0.91 mm

   **b.** 89.4 cm$^2$ × 4.8 cm

   **c.** 14.9 m$^3$ ÷ 3.0 m$^2$

   **d.** 6.975 m × 30 m × 21.5 m

9. What is the volume of a region of space that measures 752 m × 319 m × 110 m? Give your answer in the correct unit and with the proper number of significant figures.

10. Perform the following calculations, and express the results in the correct units and with the proper number of significant figures.

    **a.** 7.382 g + 1.21 g + 4.7923 g

    **b.** 51.3 mg + 83 mg − 34.2 mg

    **c.** 0.007 L − 0.0037 L + 0.012 L

    **d.** 253.05 cm$^2$ + 33.9 cm$^2$ + 28 cm$^2$

    **e.** 14.77 kg + 0.086 kg − 0.391 kg

    **f.** 319 mL + 13.75 mL + 20. mL

11. A container measures 30.5 mm × 202 mm × 153 mm. When it is full of a liquid, it has a mass of 1.33 kg. When it is empty, it has a mass of 0.30 kg. What is the density of the liquid in kilograms per liter?

12. If 7.76 km of wire has a mass of 3.3 kg, what is the mass of the wire in g/m? What length in meters would have a mass of 1.0 g?

13. A container of plant food recommends an application rate of 52 kg/ha. If the container holds 10 kg of plant food, how many square meters will it cover (1 ha = 10 000 m$^2$)?

14. A chemical process produces 974 550 kJ of energy as heat in 37.0 min. What is the rate in kilojoules per minute? What is the rate in kilojoules per second?

**| Problem Solving** *continued*

15. A water pipe fills a container that measures 189 cm × 307 cm × 272 cm in 97 s.

   **a.** What is the volume of the container in cubic meters?

   **b.** What is the rate of flow in the pipe in liters per minute?

   **c.** What is the rate of flow in cubic meters per hour?

16. Perform the following calculations, and express the results in the correct units and with the proper number of significant figures. Note, in problems with multiple steps, it is better to perform the entire calculation and then round to significant figures.

   **a.** $(0.054 \text{ kg} + 1.33 \text{ kg}) \times 5.4 \text{ m}^2$

   **b.** $67.35 \text{ cm}^2 \div (1.401 \text{ cm} - 0.399 \text{ cm})$

   **c.** $4.198 \text{ kg} \times (1019 \text{ m}^2 - 40 \text{ m}^2) \div (54.2 \text{ s} \times 31.3 \text{ s})$

   **d.** $3.14159 \text{ m} \times (4.17 \text{ m} + 2.150 \text{ m})$

   **e.** $690\,000 \text{ m} \div (5.022 \text{ h} - 4.31 \text{ h})$

   **f.** $(6.23 \text{ cm} + 3.111 \text{ cm} - 0.05 \text{ cm}) \times 14.99 \text{ cm}$

Skills Worksheet

# Problem Solving

## Scientific Notation

People who work in scientific fields often have to use very large and very small numbers. Look at some examples in the following table:

| Measurement | Value |
|---|---|
| Density of air at 27°C and 1 atm pressure | 0.001 61 g/cm$^3$ |
| Radius of a calcium atom | 0.000 000 000 197 m |
| One light-year | 9 460 000 000 000 km |
| The mass of a neutron | 0.000 000 000 000 000 000 000 001 675 g |

You can see that measurements such as these would be awkward to write out repeatedly. Also, calculating with very long numbers is likely to lead to errors because it's so easy to miscount zeros and decimal places. To make these numbers easier to handle, scientists express them in a form known as *scientific notation*, which uses powers of 10 to reduce the number of zeros to a minimum.

Look at a simple example of the way that scientific notation works. Following are some powers of 10 and their decimal equivalents.

$$10^{-2} = 0.01$$
$$10^{-1} = 0.1$$
$$10^0 = 1$$
$$10^1 = 10$$
$$10^2 = 100$$

Suppose we rewrite the values in the table using scientific notation. The numbers become much less cumbersome.

| Measurement | Value |
|---|---|
| Density of air at 27°C and 1 atm pressure | $1.61 \times 10^{-3}$ g/cm$^3$ |
| Radius of a calcium atom | $1.97 \times 10^{-10}$ m |
| One light-year | $9.46 \times 10^{12}$ km |
| Mass of a neutron | $1.675 \times 10^{-24}$ g |

**Problem Solving** *continued*

# CONVERTING QUANTITIES TO SCIENTIFIC NOTATION

**General Plan for Converting Quantities to Scientific Notation**

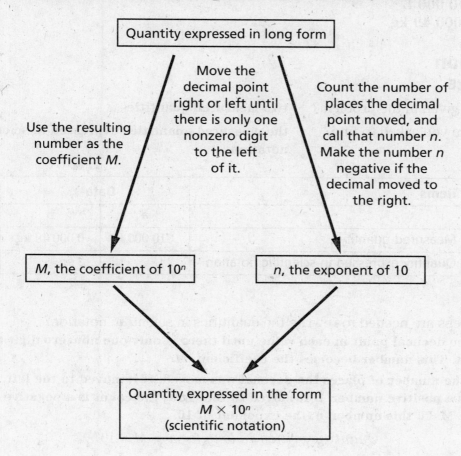

Quantity expressed in long form

Move the decimal point right or left until there is only one nonzero digit to the left of it.

Count the number of places the decimal point moved, and call that number *n*. Make the number *n* negative if the decimal moved to the right.

Use the resulting number as the coefficient *M*.

$M$, the coefficient of $10^n$

$n$, the exponent of 10

Quantity expressed in the form
$M \times 10^n$
(scientific notation)

## Sample Problem 1

**Express the following measurements in scientific notation.**
   **a. 310 000 L**
   **b. 0.000 49 kg**

## Solution

### ANALYZE

What is given in the problem?     **two measured quantities**

What are you asked to find?       **the measured quantities expressed in scientific**
                                  **notation**

| Items | Data | |
|---|---|---|
| | **a** | **b** |
| Measured quantity | 310 000 L | 0.000 49 kg |
| Quantity expressed in scientific notation | ? L | ? kg |

### PLAN

What steps are needed to rewrite the quantities in scientific notation?
**Move the decimal point in each value until there is only one nonzero digit to the
left of it. This number becomes the coefficient, M.**

**Count the number of places the decimal was moved. If it moved to the left, the
count is a positive number. If it moved to the right, the count is a negative
number. Make this number, n, the exponent of 10.**

$$Quantity\ written\ in\ long\ form = M \times 10^n$$

### COMPUTE

**a.** Express 310 000 L in scientific notation

$M = 3.1$

$3\,10\,000$    *decimal point moves
5 places to the left*
$n = +5$

$310\,000\ L = 3.1 \times 10^5\ L$

**b.** Express 0.000 49 kg in scientific notation.

$M = 4.9$

$0.00049$    *decimal point moves
4 places to the right*
$n = -4$

$0.000\,49\ kg = 4.9 \times 10^{-4}\ kg$

**Problem Solving** *continued*

## EVALUATE

|  | **a** | **b** |
|---|---|---|
| Are units correct? | Yes; the original measurement was in liters. | Yes; the original measurement was in kilograms. |
| Is the quantity correctly expressed? | Yes; the decimal was moved to the left five places to give a coefficient of 3.1 and an exponent of +5. | Yes; the decimal was moved to the right four places to give a coefficient of 4.9 and an exponent of −4. |

# Practice

**1.** Express the following quantities in scientific notation:

**a.** 8 800 000 000 m **ans: 8.8 × 10⁹ m**

**b.** 0.0015 kg **ans: 1.5 × 10⁻³ kg**

**c.** 0.000 000 000 06 kg/m³ **ans: 6 × 10⁻¹¹ kg/m³**

**d.** 8 002 000 Hz **ans: 8.002 × 10⁶ Hz**

**e.** 0.009 003 amp **ans: 9.003 × 10⁻³ amp**

**f.** 70 000 000 000 000 000 km **ans: 7 × 10¹⁶ km**

**g.** 6028 L **ans: 6.028 × 10³ L**

**h.** 0.2105 g **ans: 2.105 × 10⁻¹ g**

**i.** 600 005 000 kJ/h **ans: 6.000 05 × 10⁸ kJ/h**

**j.** 33.8 m² **ans: 3.38 × 10¹ m²**

| Problem Solving *continued*

## CALCULATING WITH QUANTITIES IN SCIENTIFIC NOTATION

## Sample Problem 2

What is the total of the measurements $3.61 \times 10^4$ mm, $5.88 \times 10^3$ mm, and $8.1 \times 10^2$ mm?

## Solution

### ANALYZE

What is given in the problem?      **three measured quantities expressed in scientific notation**

What are you asked to find?      **the sum of those quantities**

| Items | Data | | |
|---|---|---|---|
| Measured quantity | $3.61 \times 10^4$ mm | $5.88 \times 10^3$ mm | $8.1 \times 10^2$ mm |

### PLAN

What steps are needed to add the quantities?

**Convert each quantity so that each exponent is the same as that on the quantity with the largest exponent. The quantities can then be added together. Make sure the result has the correct number of significant figures.**

$$(P \times 10^Q) + (R \times 10^S) + (T \times 10^V) = ?  \qquad (P + R' + T') \times 10^Q = ?$$

*if the exponents are different, convert the quantities so that they have the same exponent as the term with the largest exponent*

*add the quantities P, R', and T', and multiply them by the factor $10^Q$*

$$(P \times 10^Q) + (R' \times 10^Q) + (T' \times 10^Q) = ?$$

### COMPUTE

$$3.61 \times 10^4 \text{ mm} + 5.88 \times 10^3 \text{ mm} + 8.1 \times 10^2 \text{ mm} = ? \text{ mm}$$

Convert the second and third quantities to multiples of $10^4$.

To convert $5.88 \times 10^3$ mm:

$$5.88 \times 10^3 \text{ mm} = M \times 10^4 \text{ mm}$$

Because one was added to the exponent, the decimal point must be moved one place to the left.

$$M = 0.588$$

To convert $8.1 \times 10^2$ mm:

$$8.1 \times 10^2 \text{ mm} = M \times 10^4 \text{ mm}$$

Because two was added to the exponent, the decimal point must be moved two places to the left.

**Problem Solving** *continued*

$$M = 0.081$$

Now the three quantities can be added, as follows:

$$
\begin{array}{r}
3.61 \ \times 10^4 \ \text{mm} \\
+ \ 0.588 \times 10^4 \ \text{mm} \\
+ \ 0.081 \times 10^4 \ \text{mm} \\
\hline
4.275 \times 10^4 \ \text{mm}
\end{array}
$$

To express the result in the correct number of significant figures, note that the result should only contain two decimal places.

$$4.27\cancel{5} \times 10^4 \ \text{mm} = 4.28 \times 10^4 \ \text{mm}$$

*this digit is 5, so round up*

## EVALUATE

Are the units correct?
**Yes; units of all quantities were millimeters.**

Is the quantity correctly expressed in scientific notation?
**Yes; there is only one number to the left of the decimal point.**

Is the quantity expressed in the correct number of significant figures?
**Yes; the result was rounded to give two decimal places to match the least accurate measurement.**

## Practice

1. Carry out the following calculations. Express the results in scientific notation and with the correct number of significant figures.

   **a.** $4.74 \times 10^4$ km + $7.71 \times 10^3$ km + $1.05 \times 10^3$ km **ans: 5.62 × 10⁴ km**

   **b.** $2.75 \times 10^{-4}$ m + $8.03 \times 10^{-5}$ m + $2.122 \times 10^{-3}$ m **ans: 2.477 × 10⁻³ m**

**c.** $4.0 \times 10^{-5} \text{ m}^3 + 6.85 \times 10^{-6} \text{ m}^3 - 1.05 \times 10^{-5} \text{ m}^3$ **ans: $3.6 \times 10^{-5}$ m$^3$**

**d.** $3.15 \times 10^2 \text{ mg} + 3.15 \times 10^3 \text{ mg} + 3.15 \times 10^4 \text{ mg}$ **ans: $3.50 \times 10^4$ mg**

**e.** $3.01 \times 10^{22} \text{ atoms} + 1.19 \times 10^{23} \text{ atoms} + 9.80 \times 10^{21} \text{ atoms}$
**ans: $1.59 \times 10^{23}$ atoms**

**f.** $6.85 \times 10^7 \text{ nm} + 4.0229 \times 10^8 \text{ nm} - 8.38 \times 10^6 \text{ nm}$ **ans: $4.624 \times 10^8$ nm**

**| Problem Solving** *continued*

# Sample Problem 3

**Perform the following calculation, and express the result in scientific notation:**

$$3.03 \times 10^4 \text{ cm}^2 \times 6.29 \times 10^2 \text{ cm}$$

# Solution

## ANALYZE

What is given in the problem?       **two quantities expressed in scientific notation**

What are you asked to find?       **the product of the two quantities**

| Items | Data | |
|---|---|---|
| Measured quantity | $3.03 \times 10^4$ cm$^2$ | $6.29 \times 10^2$ cm |

## PLAN

What steps are needed to multiply quantities expressed in scientific notation?
**Multiply the coefficients, and add the exponents. Then transform to the correct scientific notation form with the correct units and number of significant figures.**

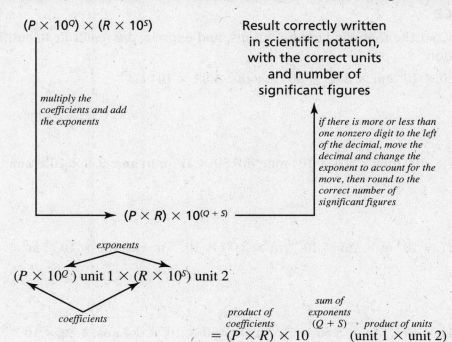

## COMPUTE

$3.03 \times 10^4 \text{ cm}^2 \times 6.29 \times 10^2 \text{ cm} =$

$$(3.03 \times 6.29) \times 10^{(4+2)} \text{ (cm}^2 \times \text{cm)} = 19.0587 \times 10^6 \text{ cm}^3$$

To transform the result to the correct form for scientific notation, move the decimal point left one place and increase the exponent by one.

$$19.0587 \times 10^6 \text{ cm}^3 = 1.90587 \times 10^{(6+1)} \text{ cm}^3 = 1.90587 \times 10^7 \text{ cm}^3$$

To express the result to the correct number of significant figures, note that both of the original quantities have three significant figures. Therefore, round off the result to three significant figures.

$$1.905\,87 \times 10^7 \text{ cm}^3 = 1.91 \times 10^7 \text{ cm}^3$$

*this digit is 5,*
*so round up*

### EVALUATE

Are the units correct?
**Yes; the units cm$^2$ and cm are multiplied to give cm$^3$.**

Is the quantity expressed to the correct number of significant figures?
**Yes; the number of significant figures is correct because the data were given to three significant figures.**

Is the quantity expressed correctly in scientific notation?
**Yes; moving the decimal point decreases the coefficient by a factor of 10, so the exponent increases by one to compensate.**

## Practice

**1.** Carry out the following computations, and express the result in scientific notation:

  **a.** $7.20 \times 10^3$ cm $\times$ $8.08 \times 10^3$ cm **ans: 5.82 $\times$ 10$^7$ cm$^2$**

  **b.** $3.7 \times 10^4$ mm $\times$ $6.6 \times 10^4$ mm $\times$ $9.89 \times 10^3$ mm **ans: 2.4 $\times$ 10$^{13}$ mm$^3$**

  **c.** $8.27 \times 10^2$ m $\times$ $2.5 \times 10^{-3}$ m $\times$ $3.00 \times 10^{-4}$ m **ans: 6.2 $\times$ 10$^{-4}$ m$^3$**

  **d.** $4.44 \times 10^{-35}$ m $\times$ $5.55 \times 10^{19}$ m $\times$ $7.69 \times 10^{-12}$ kg **ans: 1.89 $\times$ 10$^{-26}$ kg·m$^2$**

  **e.** $6.55 \times 10^4$ dm $\times$ $7.89 \times 10^9$ dm $\times$ $4.01893 \times 10^5$ dm **ans: 2.08 $\times$ 10$^{20}$ dm$^3$**

**Problem Solving** *continued*

# Sample Problem 4

**Perform the following calculation, and express the result in scientific notation:**

$$3.803 \times 10^3 \text{ g} \div 5.3 \times 10^6 \text{ mL}$$

# Solution

## ANALYZE

What is given in the problem?     **two quantities expressed in scientific notation**

What are you asked to find?     **the quotient of the two quantities**

| Items | Data | |
|---|---|---|
| Measured quantity | $3.803 \times 10^3$ g | $5.3 \times 10^6$ mL |

## PLAN

What steps are needed to divide the quantities expressed in scientific notation?
**Divide the coefficients, and subtract the exponents. Then transform the result
to the correct form for scientific notation with the correct units and number of
significant figures.**

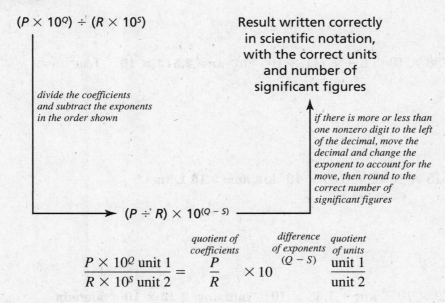

Result written correctly
in scientific notation,
with the correct units
and number of
significant figures

if there is more or less than
one nonzero digit to the left
of the decimal, move the
decimal and change the
exponent to account for the
move, then round to the
correct number of
significant figures

$$\frac{P \times 10^Q \text{ unit 1}}{R \times 10^S \text{ unit 2}} = \frac{P}{R} \times 10^{(Q-S)} \frac{\text{unit 1}}{\text{unit 2}}$$

## COMPUTE

$$3.803 \times 10^3 \text{ g} \div 5.3 \times 10^6 \text{ mL} = \frac{3.803}{5.3} \times 10^{(3-6)} \frac{\text{g}}{\text{mL}} = 0.717\ 547 \times 10^{-3} \text{ g/mL}$$

The measurement $5.3 \times 10^6$ mL has the fewest significant figures; round the
result accordingly.

$$0.717\ 547 \times 10^{-3} \text{ g/mL} = 0.72 \times 10^{-3} \text{ g/mL}$$

*this digit is greater
than 5, so round up*

**| Problem Solving** *continued*

To transform the result to the correct form for scientific notation, move the decimal point to the right one place and decrease the exponent by one.

$$0.72 \times 10^{-3} \text{ g/mL} = 7.2 \times 10^{(-3-1)} \text{ g/mL} = 7.2 \times 10^{-4} \text{ g/mL}$$

## EVALUATE
Are the units correct?
**Yes; grams divided by milliliters gives g/mL, a unit of density.**

Is the quantity expressed to the correct number of significant figures?
**Yes; the result was limited to two significant figures by the data given.**

Is the quantity expressed correctly in scientific notation?
**Yes; there is only one nonzero digit to the left of the decimal point.**

# Practice

1. Carry out the following computations, and express the result in scientific notation:

   **a.** $2.290 \times 10^7$ cm ÷ $4.33 \times 10^3$ s **ans: $5.29 \times 10^3$ cm/s**

   **b.** $1.788 \times 10^{-5}$ L ÷ $7.111 \times 10^{-3}$ m$^2$ **ans: $2.514 \times 10^{-3}$ L/m$^2$**

   **c.** $5.515 \times 10^4$ L ÷ $6.04 \times 10^3$ km **ans: $9.13$ L/km**

   **d.** $3.29 \times 10^{-4}$ km ÷ $1.48 \times 10^{-2}$ min **ans: $2.22 \times 10^{-2}$ km/min**

   **e.** $4.73 \times 10^{-4}$ g ÷ $(2.08 \times 10^{-3}$ km × $5.60 \times 10^{-4}$ km) **ans: $4.06 \times 10^2$ g/km$^2$**

| **Problem Solving** *continued* |
| --- |

# Additional Problems

1. Express the following quantities in scientific notation:

   **a.** 158 000 km

   **b.** 0.000 009 782 L

   **c.** 837 100 000 cm$^3$

   **d.** 6 500 000 000 mm$^2$

   **e.** 0.005 93 g

   **f.** 0.000 000 006 13 m

   **g.** 12 552 000 J

   **h.** 0.000 008 004 g/L

   **i.** 0.010 995 kg

   **j.** 1 050 000 000 Hz

2. Perform the following calculations, and express the result in scientific notation with the correct number of significant figures:

   **a.** $2.48 \times 10^2$ kg + $9.17 \times 10^3$ kg + $7.2 \times 10^1$ kg

   **b.** $4.07 \times 10^{-5}$ mg + $3.966 \times 10^{-4}$ mg + $7.1 \times 10^{-2}$ mg

   **c.** $1.39 \times 10^4$ m$^3$ + $6.52 \times 10^2$ m$^3$ − $4.8 \times 10^3$ m$^3$

   **d.** $7.70 \times 10^{-9}$ m − $3.95 \times 10^{-8}$ m + $1.88 \times 10^{-7}$ m

   **e.** $1.111 \times 10^5$ J + $5.82 \times 10^4$ J + $3.01 \times 10^6$ J

   **f.** $9.81 \times 10^{27}$ molecules + $3.18 \times 10^{25}$ molecules − $2.09 \times 10^{26}$ molecules

   **g.** $1.36 \times 10^7$ cm + $3.456 \times 10^6$ cm − $1.01 \times 10^7$ cm + $5.122 \times 10^5$ cm

3. Perform the following computations, and express the result in scientific notation with the correct number of significant figures:

   **a.** $1.54 \times 10^{-1}$ L ÷ $2.36 \times 10^{-4}$ s

   **b.** $3.890 \times 10^4$ mm × $4.71 \times 10^2$ mm$^2$

   **c.** $9.571 \times 10^3$ kg ÷ $3.82 \times 10^{-1}$ m$^2$

   **d.** $8.33 \times 10^3$ km ÷ $1.97 \times 10^2$ s

   **e.** $9.36 \times 10^2$ m × $3.82 \times 10^3$ m × $9.01 \times 10^{-1}$ m

   **f.** $6.377 \times 10^4$ J ÷ $7.35 \times 10^{-3}$ s

4. Your electric company charges you for the electric energy you use, measured in kilowatt-hours (kWh). One kWh is equivalent to 3 600 000 J. Express this quantity in scientific notation.

5. The pressure in the deepest part of the ocean is 11 200 000 Pa. Express this pressure in scientific notation.

6. Convert 1.5 km to millimeters, and express the result in scientific notation.

7. Light travels at a speed of about 300 000 km/s.

   **a.** Express this value in scientific notation.

   **b.** Convert this value to meters per hour.

   **c.** What distance in centimeters does light travel in 1 μs?

8. There are $7.11 \times 10^{24}$ molecules in 100.0 cm$^3$ of a certain substance.

   **a.** What is the number of molecules in 1.09 cm$^3$ of the substance?

   **b.** What would be the number of molecules in $2.24 \times 10^4$ cm$^3$ of the substance?

   **c.** What number of molecules are in $9.01 \times 10^{-6}$ cm$^3$ of the substance?

9. The number of transistors on a particular integrated circuit is 3 578 000, and the integrated circuit measures 9.5 mm × 8.2 mm.

   **a.** What is the area occupied by each transistor?

   **b.** Using your answer from (a), how many transistors could be formed on a silicon sheet that measures 353 mm × 265 mm?

10. A solution has 0.0501 g of a substance in 1.00 L. Express this concentration in grams per microliter.

11. Cesium atoms are the largest of the naturally occurring elements. They have a diameter of $5.30 \times 10^{-10}$ m. Calculate the number of cesium atoms that would have to be lined up to give a row of cesium atoms 2.54 cm (1 in.) long.

12. The neutron has a volume of approximately $1.4 \times 10^{-44}$ m$^3$ and a mass of $1.675 \times 10^{-24}$ g. Calculate the density of the neutron in g/m$^3$. What is the mass of 1.0 cm$^3$ of neutrons in kilograms?

13. The pits in a compact disc are some of the smallest things ever mass-produced mechanically by humans. These pits represent the *1*s and *0*s of digital information on a compact disc. These pits are only $1.6 \times 10^{-8}$ m deep (1/4 the wavelength of red laser light). How many of these pits would have to be stacked on top of each other to make a hole 0.305 m deep?

14. 22 400 mL of oxygen gas contains $6.022 \times 10^{23}$ oxygen molecules at 0°C and standard atmospheric pressure.

   **a.** How many oxygen molecules are in 0.100 mL of gas?

   **b.** How many oxygen molecules are in 1.00 L of gas?

   **c.** What is the average space in milliters occupied by one oxygen molecule?

15. The mass of the atmosphere is calculated to be $5.136 \times 10^{18}$ kg, and there are 6 500 000 000 people living on Earth. Calculate the following values.

   **a.** The mass of atmosphere in kilograms per person.

   **b.** The mass of atmosphere in metric tons per person.

   **c.** If the number of people increases to 9 500 000 000, what is the mass in kilograms per person?

16. The mass of the sun is $1.989 \times 10^{30}$ kg, and the mass of Earth is $5.974 \times 10^{24}$ kilograms. How many Earths would be needed to equal the mass of the sun?

| Problem Solving *continued*

**17.** A new landfill has dimensions of 2.3 km $\times$ 1.4 km $\times$ 0.15 km.

   **a.** What is the volume in cubic kilometer?

   **b.** What is the volume in cubic meters?

   **c.** If 250 000 000 objects averaging 0.060 m$^3$ each are placed into the landfill each year, how many years will it take to fill the landfill?

**18.** A dietary calorie (C) is exactly equal to 1000 cal. If your daily intake of food gives you 2400 C, what is your intake in joules per day? (1 cal = 4.184 J)

Skills Worksheet )

# Problem Solving

## Four Steps for Solving Quantitative Problems

Maybe you have noticed that the sample problems in the first three chapters of this book are solved in a four-step process. The steps in the process are as follows.

1. **ANALYZE**
2. **PLAN**
3. **COMPUTE**
4. **EVALUATE**

Now it's time to examine each step to learn how this process can help you solve more-complex problems, like those you will encounter in your chemistry course.

## 1. ANALYZE

In this first step, you should read the problem carefully and then reread it. You must determine what specific information and data you are given in the problem and what you need to find.

Try to visualize the situation the problem describes. Look closely at the words in the problem statement for clues to understanding the problem. Are you working with elements, compounds, or mixtures? Are they solids, liquids, or gases? What change is taking place? Is it a chemical change? What are the reactants? What are the products?

It is always a good idea to collect and organize all of your information in a table, where you can see it at a glance. Include the things you want to find in the table, too. Be sure to include units with both the data and the quantities you must find. Scanning the quantities and their units will often provide clues about how to set up the problem. Remember, you may be given information not needed to solve the problem. You must analyze the data to determine what is useful and what is not.

## 2. PLAN

In the planning step, you develop a method to solve the problem. Always keep in mind what you want to find and its units. Chances are good that an approach that gives an answer with the correct units is the correct one to use. In any case, a setup that gives an answer with the wrong units is certain to be wrong. You may find it helpful to diagram your solution method. This process helps you organize your thoughts.

As you work out your problem-solving method, write down a trial calculation without numbers but with units. When you complete your setup, see if the trial calculation will give you a quantity with the correct units. As stated above, if the setup gives the needed units, it is probably correct.

**| Problem Solving** *continued*

During this planning process you may discover that you need more information, such as atomic masses from the periodic table, the boiling point of alcohol, or the density of tin. You will need to look up such information in the appropriate tables.

## 3. COMPUTE

In this step, you follow your plan, set up a calculation using the data you have assembled, and compute the result.

It is a good strategy to write out and check your calculation setup before you start working with your calculator. First reconfirm that the calculation will give a result with the correct units. Go through your calculation and lightly strike through the units that cancel. Be sure the remaining units are those that you want in your answer. Whenever possible, use your calculator in a way that lets you complete the entire problem without writing down numbers and then re-entering them.

## 4. EVALUATE

Everyone makes errors, but good problem solvers always develop strategies to check their work. Confidence in your problem-solving ability will come from knowing how to determine on your own whether your answers are correct.

One evaluation strategy is to estimate the numerical value of the answer. In simple problems, you can probably do this in your head. With calculations having several terms, round off each numerical value to the nearest simple value, and then write and compute the estimation.

Suppose you had to make the following calculation.

$$\frac{28.8 \text{ g}}{6.30 \text{ cm} \times 18.9 \text{ cm}} = ?$$

The numerical calculation can be estimated as follows.

$$\frac{30}{6 \times 20} = \frac{30}{120} = \frac{1}{4} = 0.25$$

Once you have done the actual computation, compare your result with the estimate. In this case, the calculation gives $0.242 \text{ g/cm}^2$, which is close to the estimated result. Therefore, it is likely that you made no mistakes in the calculation.

Next, check that your answer is expressed to the correct number of significant figures. Look at the data values you used in the calculation. Usually, significant figures will be limited by the measurement that has the fewest significant figures.

Finally, ask yourself the simple question, does this answer make sense based on what you know? If you are calculating the circumference of Earth, an answer of 50 km is obviously much too small. If you are calculating the density of air, a value of $340 \text{ g/cm}^3$ is much too large because air density is usually less than 1 $\text{g/cm}^3$. You will detect many errors by asking if the answer makes sense.

# Sample Problem 1

**A 10.0% sodium hydroxide solution has a density of 1.11 g/mL. What volume in liters will 2280 g of the solution have?**

# Solution

## ANALYZE

What is given in the problem?    **the density of the sodium hydroxide solution in g/mL, and the mass of the solution whose volume is to be determined**

What are you asked to find?    **the volume of the specified mass of solution**

Next, bring together in a table everything you might need in the problem. Include what you want to find in the table. The fact that the solution is 10.0% sodium hydroxide is unimportant in the solution of this problem. This piece of data need not appear in your table. Notice that the problem asks you for a volume in liters and that density is given in grams per milliliter. You will need a factor to convert between these units. Therefore, include in the table any relationships between units that will be helpful.

| Items | Data |
|---|---|
| Density of solution | 1.11 g/mL |
| Mass of solution | 2280 g |
| Volume of solution | ? L |
| Relationship between mL and L | 1000 mL = 1 L |

## PLAN

What steps are needed to calculate the volume of 2280 g of the solution?
**Apply the relationship $D = \dfrac{m}{V}$. Rearrange to solve for V, substitute data, and convert to liters.**

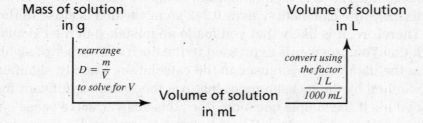

Solve the density equation for V.

$$V = \frac{m}{D}$$

$$\frac{\text{mass of solution in g}}{\text{density of solution in g/ml}} = \text{volume in mL}$$

To change the result to liters, multiply by the conversion factor.

$$\frac{\text{mass of solution in g}}{\text{density of solution g/ml}} \times \frac{1\ L}{1000\ mL} = \text{volume in L}$$

**COMPUTE**

$$\frac{2280\ g}{1.11\ g/mL} \times \frac{1\ L}{1000\ mL} = 2.05\ L$$

**EVALUATE**

Are the units correct?

**Yes; units canceled to give liters.**

Is the number of significant figures correct?

**Yes; the number of significant figures is correct because data were given to three significant figures.**

Is the answer reasonable?

**Yes; the calculation can be approximated as 2000/1000 = 2. Also, considering that 2000 g of water occupies 2 L, you would expect 2 L of a slightly more dense material to have a mass slightly greater than 2000 g.**

# Practice

1. Gasoline has a density of 0.73 g/cm$^3$. How many liters of gasoline would be required to increase the mass of an automobile from 1271 kg to 1305 kg?
   **ans: 47 L**

2. A swimming pool measures 9.0 m long by 3.5 m wide by 1.75 m deep. What mass of water in metric tons (1 metric ton = 1000 kg) does the pool contain when filled? The density of the water in the pool is 0.997 g/cm$^3$.
   **ans: 55 metric tons**

3. A tightly packed box of crackers contains 250 g of crackers and measures 7.0 cm × 17.0 cm × 19.0 cm. What is the average density in kilograms per liter of the crackers in the package? Assume that the unused volume is negligible.
   **ans: 0.11 kg/L**

## Additional Problems

**Solve these problems by using the Four Steps for Solving Quantitative Problems.**

**1.** The aluminum foil on a certain roll has a total area of 18.5 m$^2$ and a mass of 1275 g. Using a density of 2.7 g per cubic centimeter for aluminum, determine the thickness in millimeters of the aluminum foil.

**2.** If a liquid has a density of 1.17 g/cm$^3$, how many liters of the liquid have a mass of 3.75 kg?

**3.** A stack of 500 sheets of paper measuring 28 cm $\times$ 21 cm is 44.5 mm high and has a mass of 2090 g. What is the density of the paper in grams per cubic centimeter?

**4.** A triangular-shaped piece of a metal has a mass of 6.58 g. The triangle is 0.560 mm thick and measures 36.4 mm on the base and 30.1 mm in height. What is the density of the metal in grams per cubic centimeter?

**5.** A packing crate measures 0.40 m $\times$ 0.40 m $\times$ 0.25 m. You must fill the crate with boxes of cookies that each measure 22.0 cm $\times$ 12.0 cm $\times$ 5.0 cm. How many boxes of cookies can fit into the crate?

**6.** Calculate the unknown quantities in the following table. Use the following relationships for volumes of the various shapes.

$$\text{Volume of a cube} = l \times l \times l$$
$$\text{Volume of a rectangle} = l \times w \times h$$
$$\text{Volume of a sphere} = 4/3\pi r^3$$
$$\text{Volume of a cylinder} = \pi r^2 \times h$$

|     | *D*             | *m*      | *V*        | **Shape** | **Dimensions**                              |
|-----|-----------------|----------|------------|-----------|---------------------------------------------|
| **a.** | 2.27 g/cm$^3$ | 3.93 kg  | ? L        | cube      | ? m $\times$ ? m $\times$ ? m               |
| **b.** | 1.85 g/cm$^3$ | ? g      | ? cm$^3$   | rectangle | 33 mm $\times$ 21 mm $\times$ 7.2 mm        |
| **c.** | 3.21 g/L      | ? kg     | ? dm$^3$   | sphere    | 3.30 m diameter                             |
| **d.** | ? g/cm$^3$    | 497 g    | ? m$^3$    | cylinder  | 7.5 cm diameter $\times$ 12 cm              |
| **e.** | 0.92 g/cm$^3$ | ? kg     | ? cm$^3$   | rectangle | 3.5 m $\times$ 1.2 m $\times$ 0.65 m        |

**7.** When a sample of a metal alloy that has a mass of 9.65 g is placed into a graduated cylinder containing water, the volume reading in the cylinder increases from 16.0 mL to 19.5 mL. What is the density of the alloy sample in grams per cubic centimeter?

**8.** Pure gold can be made into extremely thin sheets called gold leaf. Suppose that 50.0 kg of gold is made into gold leaf having an area of 3620 m$^2$. The density of gold is 19.3 g/cm$^3$.

**a.** How thick in micrometers is the gold leaf?

**b.** A gold atom has a radius of $1.44 \times 10^{-10}$ m. How many atoms thick is the gold leaf?

**Problem Solving** *continued*

9. A chemical plant process requires that a cylindrical reaction tank be filled with a certain liquid in 238 s. The tank is 1.2 m in diameter and 4.6 m high. What flow rate in liters per minute is required to fill the reaction tank in the specified time?

10. The radioactive decay of 2.8 g of plutonium-238 generates 1.0 joule of energy as heat every second. Plutonium has a density of 19.86 g/cm$^3$. How many calories (1 cal = 4.184 J) of energy as heat will a rectangular piece of plutonium that is 4.5 cm × 3.05 cm × 15 cm generate per hour?

11. The mass of Earth is 5.974 × 10$^{24}$ kg. Assume that Earth is a sphere of diameter 1.28 × 10$^4$ km and calculate the average density of Earth in grams per cubic centimeter.

12. What volume of magnesium in cubic centimeters would have the same mass as 1.82 dm$^3$ of platinum? The density of magnesium is 1.74 g/cm$^3$, and the density of platinum is 21.45 g/cm$^3$.

13. A roll of transparent tape has 66 m of tape on it. If an average of 5.0 cm of tape is needed each time the tape is used, how many uses can you get from a case of tape containing 24 rolls?

14. An automobile can travel 38 km on 4.0 L of gasoline. If the automobile is driven 75% of the days in a year and the average distance traveled each day is 86 km, how many liters of gasoline will be consumed in one year (assume the year has 365 days)?

15. A hose delivers water to a swimming pool that measures 9.0 m long by 3.5 m wide by 1.75 m deep. It requires 97 h to fill the pool. At what rate in liters per minute will the hose fill the pool?

16. Automobile batteries are filled with a solution of sulfuric acid, which has a density of 1.285 g/cm$^3$. The solution used to fill the battery is 38% (by mass) sulfuric acid. How many grams of sulfuric acid are present in 500 mL of battery acid?

Skills Worksheet

# Problem Solving

## Mole Concept

Suppose you want to carry out a reaction that requires combining one atom of iron with one atom of sulfur. How much iron should you use? How much sulfur? When you look around the lab, there is no device that can count numbers of atoms. Besides, the merest speck (0.001 g) of iron contains over a billion billion atoms. The same is true of sulfur.

Fortunately, you do have a way to relate mass and numbers of atoms. One iron atom has a mass of 55.847 amu, and 55.847 g of iron contains $6.022\ 137 \times 10^{23}$ atoms of iron. Likewise, 32.066 g of sulfur contains $6.022\ 137 \times 10^{23}$ atoms of sulfur. Knowing this, you can measure out 55.847 g of iron and 32.066 g of sulfur and be pretty certain that you have the same number of atoms of each.

The number $6.022\ 137 \times 10^{23}$ is called Avogadro's number. For most purposes it is rounded off to $6.022 \times 10^{23}$. Because this is an awkward number to write over and over again, chemists refer to it as a *mole* (abbreviated mol). $6.022 \times 10^{23}$ objects is called a mole, just as you call 12 objects a dozen.

Look again at how these quantities are related.

$$55.847 \text{ g of iron} = 6.022 \times 10^{23} \text{ iron atoms} = 1 \text{ mol of iron}$$

$$32.066 \text{ g of sulfur} = 6.022 \times 10^{23} \text{ sulfur atoms} = 1 \text{ mol of sulfur}$$

**General Plan for Converting Mass, Amount, and Numbers of Particles**

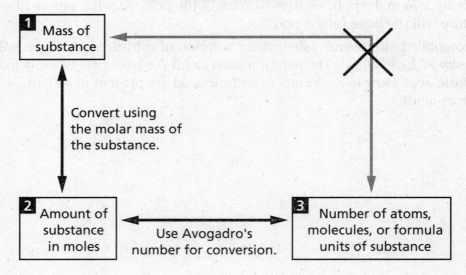

# PROBLEMS INVOLVING ATOMS AND ELEMENTS

## Sample Problem 1

A chemist has a jar containing 388.2 g of iron filings. How many moles of iron does the jar contain?

## Solution

### ANALYZE

What is given in the problem?  **mass of iron in grams**

What are you asked to find?  **amount of iron in moles**

| Items | Data |
|-------|------|
| Mass of iron | 388.2 g |
| Molar mass of iron* | 55.85 g/mol |
| Amount of iron | ? mol |

\* determined from the periodic table

### PLAN

What step is needed to convert from grams of Fe to number of moles of Fe?
**The molar mass of iron can be used to convert mass of iron to amount of iron in moles.**

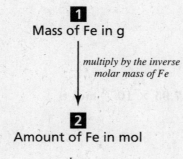

**1**
Mass of Fe in g

*multiply by the inverse molar mass of Fe*

**2**
Amount of Fe in mol

$$\text{g Fe} \times \overset{\dfrac{1}{molar\ mass\ Fe}}{\overbrace{\dfrac{1\ \text{mol Fe}}{55.85\ \text{g Fe}}}} = \text{mol Fe}$$

### COMPUTE

$$388.2\ \text{g Fe} \times \frac{1\ \text{mol Fe}}{55.85\ \text{g Fe}} = 6.951\ \text{mol Fe}$$

### EVALUATE

Are the units correct?
**Yes; the answer has the correct units of moles of Fe.**

**| Problem Solving** *continued*

Is the number of significant figures correct?

**Yes; the number of significant figures is correct because there are four significant figures in the given value of 388.2 g Fe.**

Is the answer reasonable?

**Yes; 388.2 g Fe is about seven times the molar mass. Therefore, the sample contains about 7 mol.**

## Practice

1. Calculate the number of moles in each of the following masses:

   **a.** 64.1 g of aluminum **ans: 2.38 mol Al**

   **b.** 28.1 g of silicon **ans: 1.00 mol Si**

   **c.** 0.255 g of sulfur **ans: $7.95 \times 10^{-3}$ mol S**

   **d.** 850.5 g of zinc **ans: 13.01 mol Zn**

**| Problem Solving** *continued*

## Sample Problem 2

**A student needs 0.366 mol of zinc for a reaction. What mass of zinc in grams should the student obtain?**

## Solution

### ANALYZE

What is given in the problem?     **amount of zinc needed in moles**

What are you asked to find?     **mass of zinc in grams**

| Items | Data |
|---|---|
| Amount of zinc | 0.366 mol |
| Molar mass of zinc | 65.39 g/mol |
| Mass of zinc | ? g |

### PLAN

What step is needed to convert from moles of Zn to grams of Zn?
**The molar mass of zinc can be used to convert amount of zinc to mass of zinc.**

$$\boxed{2}$$
Amount of Zn in mol $\xrightarrow[\substack{\text{multiply by the} \\ \text{molar mass of Zn}}]{}$ $\boxed{1}$ Mass of Zn in mol

$$\overset{given}{\text{mol Zn}} \times \frac{\overset{molar\ mass\ Zn}{65.39 \text{ g Zn}}}{1 \text{ mol Zn}} = \text{g Zn}$$

### COMPUTE

$$0.366 \text{ mol Zn} \times \frac{65.39 \text{ g Zn}}{1 \text{ mol Zn}} = 23.9 \text{ g Zn}$$

### EVALUATE

Are the units correct?
**Yes; the answer has the correct units of grams of Zn.**

Is the number of significant figures correct?
**Yes; the number of significant figures is correct because there are three significant figures in the given value of 0.366 mol Zn.**

Is the answer reasonable?
**Yes; 0.366 mol is about 1/3 mol. 23.9 g is about 1/3 the molar mass of Zn.**

**| Problem Solving** *continued*

## Practice

**1.** Calculate the mass of each of the following amounts:

**a.** 1.22 mol sodium **ans: 28.0 g Na**

**b.** 14.5 mol copper **ans: 921 g Cu**

**c.** 0.275 mol mercury **ans: 55.2 g Hg**

**d.** $9.37 \times 10^{-3}$ mol magnesium **ans: 0.228 Mg**

# Sample Problem 3

How many moles of lithium are there in $1.204 \times 10^{24}$ lithium atoms?

# Solution

## ANALYZE

What is given in the problem?     **number of lithium atoms**

What are you asked to find?     **amount of lithium in moles**

| Items | Data |
|---|---|
| Number of lithium atoms | $1.204 \times 10^{24}$ atoms |
| Avogadro's number—the number of atoms per mole | $6.022 \times 10^{23}$ atoms/mol |
| Amount of lithium | ? mol |

## PLAN

What step is needed to convert from number of atoms of Li to moles of Li?
**Avogadro's number is the number of atoms per mole of lithium and can be used to calculate the number of moles from the number of atoms.**

$$\boxed{3} \text{ Number of Li atoms} \xrightarrow[\substack{multiply\ by\ the\ inverse\ of \\ Avogadro's\ number}]{} \boxed{2} \text{ Amount of Li in mol}$$

$$\underset{given}{\text{atoms Li}} \times \overset{\overset{\textit{1}}{\overline{\textit{Avogadro's number}}}}{\frac{1 \text{ mol Li}}{6.022 \times 10^{23} \text{ atoms Li}}} = \text{mol Li}$$

## COMPUTE

$$1.204 \times 10^{24} \text{ atoms Li} \times \frac{1 \text{ mol Li}}{6.022 \times 10^{23} \text{ atoms Li}} = 1.999 \text{ mol Li}$$

## EVALUATE

Are the units correct?
**Yes; the answer has the correct units of moles of Li.**

Is the number of significant figures correct?
**Yes; four significant figures is correct.**

Is the answer reasonable?
**Yes; $1.204 \times 10^{24}$ is approximately twice Avogadro's number. Therefore, it is reasonable that this number of atoms would equal about 2 mol.**

**❙ Problem Solving** *continued*

## Practice

**1.** Calculate the amount in moles in each of the following quantities:

**a.** $3.01 \times 10^{23}$ atoms of rubidium **ans: 0.500 mol Rb**

**b.** $8.08 \times 10^{22}$ atoms of krypton **ans: 0.134 mol Kr**

**c.** 5 700 000 000 atoms of lead **ans: $9.5 \times 10^{-15}$ mol Pb**

**d.** $2.997 \times 10^{25}$ atoms of vanadium **ans: 49.77 mol V**

**Problem Solving** *continued*

## CONVERTING THE AMOUNT OF AN ELEMENT IN MOLES
## TO THE NUMBER OF ATOMS

In Sample Problem 3, you were asked to determine the number of moles in $1.204 \times 10^{24}$ atoms of lithium. Had you been given the amount in moles and asked to calculate the number of atoms, you would have simply multiplied by Avogadro's number. Steps 2 and 3 of the plan for solving Sample Problem 3 would have been reversed.

# Practice

**1.** Calculate the number of atoms in each of the following amounts:

**a.** 1.004 mol bismuth **ans: $6.046 \times 10^{23}$ atoms Bi**

**b.** 2.5 mol manganese **ans: $1.5 \times 10^{24}$ atoms Mg**

**c.** 0.000 000 2 mol helium **ans: $1 \times 10^{17}$ atoms He**

**d.** 32.6 mol strontium **ans: $1.96 \times 10^{25}$ atoms Sr**

**Problem Solving** *continued*

## Sample Problem 4

How many boron atoms are there in 2.00 g of boron?

## Solution

### ANALYZE

What is given in the problem?  **mass of boron in grams**

What are you asked to find?  **number of boron atoms**

| Items | Data |
|---|---|
| Mass of boron | 2.00 g |
| Molar mass of boron | 10.81 g/mol |
| Avogadro's number—the number of boron atoms per mole of boron | $6.022 \times 10^{23}$ atoms/mol |
| Number of boron atoms | ? atoms |

### PLAN

What steps are needed to convert from grams of B to number of atoms of B?
**First, you must convert the mass of boron to moles of boron by using the molar mass of boron. Then you can use Avogadro's number to convert amount in moles to number of atoms of boron.**

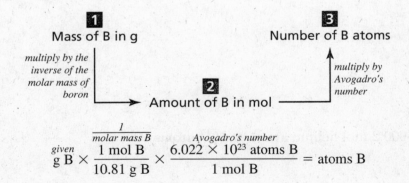

$$\overset{given}{g\ B} \times \overset{\frac{1}{molar\ mass\ B}}{\frac{1\ mol\ B}{10.81\ g\ B}} \times \overset{Avogadro's\ number}{\frac{6.022 \times 10^{23}\ atoms\ B}{1\ mol\ B}} = atoms\ B$$

### COMPUTE

$$2.00\ g\ B \times \frac{1\ mol\ B}{10.81\ g\ B} \times \frac{6.022 \times 10^{23}\ atoms\ B}{1\ mol\ B} = 1.11 \times 10^{23}\ atoms\ B$$

### EVALUATE

Are the units correct?
**Yes; the answer has the correct units of atoms of boron.**

Is the number of significant figures correct?
**Yes; the mass of boron was given to three significant figures.**

**Problem Solving** *continued*

Is the answer reasonable?
**Yes; 2 g of boron is about 1/5 of the molar mass of boron. Therefore, 2.00 g boron will contain about 1/5 of an Avogadro's constant of atoms.**

## Practice

1. Calculate the number of atoms in each of the following masses:

    **a.** 54.0 g of aluminum **ans: $1.21 \times 10^{24}$ atoms Al**

    **b.** 69.45 g of lanthanum **ans: $3.011 \times 10^{23}$ atoms La**

    **c.** 0.697 g of gallium **ans: $6.02 \times 10^{21}$ atoms Ga**

    **d.** 0.000 000 020 g beryllium **ans: $1.3 \times 10^{15}$ atoms Be**

**▌Problem Solving** *continued*

## CONVERTING NUMBER OF ATOMS OF AN ELEMENT TO MASS

Sample Problem 4 uses the progression of steps $1 \rightarrow 2 \rightarrow 3$ to convert from the mass of an element to the number of atoms. In order to calculate the mass from a given number of atoms, these steps will be reversed. The number of moles in the sample will be calculated. Then this value will be converted to the mass in grams.

## Practice

1. Calculate the mass of the following numbers of atoms:

   **a.** $6.022 \times 10^{24}$ atoms of tantalum **ans: 1810. g Ta**

   **b.** $3.01 \times 10^{21}$ atoms of cobalt **ans: 0.295 g Co**

   **c.** $1.506 \times 10^{24}$ atoms of argon **ans: 99.91 g Ar**

   **d.** $1.20 \times 10^{25}$ atoms of helium **ans: 79.7 g He**

**▌Problem Solving** *continued*

# PROBLEMS INVOLVING MOLECULES, FORMULA UNITS, AND IONS

How many water molecules are there in 200.0 g of water? What is the mass of 15.7 mol of nitrogen gas? Both of these substances consist of molecules, not single atoms. Look back at the diagram of the General Plan for Converting Mass, Amount, and Numbers of Particles. You can see that the same conversion methods can be used with molecular compounds and elements, such as $CO_2$, $H_2O$, $H_2SO_4$, and $O_2$.

For example, 1 mol of water contains $6.022 \times 10^{23}$ $H_2O$ *molecules*. The mass of a molecule of water is the sum of the masses of two hydrogen atoms and one oxygen atom, and is equal to 18.02 amu. Therefore, 1 mol of water has a mass of 18.02 g. In the same way, you can relate amount, mass, and number of formula units for ionic compounds, such as NaCl, $CaBr_2$, and $Al_2(SO_4)_3$.

## Sample Problem 5

**How many moles of carbon dioxide are in 66.0 g of dry ice, which is solid $CO_2$?**

## Solution

### ANALYZE

What is given in the problem?   **mass of carbon dioxide**

What are you asked to find?   **amount of carbon dioxide**

| Items | Data |
|---|---|
| Mass of $CO_2$ | 66.0 g |
| Molar mass of $CO_2$ | 44.0 g/mol |
| Amount of $CO_2$ | ? mol |

### PLAN

What step is needed to convert from grams of $CO_2$ to moles of $CO_2$?

**The molar mass of $CO_2$ can be used to convert mass of $CO_2$ to moles of $CO_2$.**

**1** Mass of $CO_2$ in g $\xrightarrow[\substack{\textit{multiply by the inverse of the} \\ \textit{molar mass of } CO_2}]{}$ **2** Amount of $CO_2$ in mol

$$\overset{given}{g\ CO_2} \times \overset{\frac{1}{molar\ mass\ CO_2}}{\frac{1\ \text{mol}\ CO_2}{44.01\ \text{g}\ CO_2}} = \text{mol}\ CO_2$$

### COMPUTE

$$66.0\ \text{g}\ CO_2 \times \frac{1\ \text{mol}\ CO_2}{44.01\ \text{g}\ CO_2} = 1.50\ \text{mol}\ CO_2$$

**| Problem Solving** *continued*

## EVALUATE

Are the units correct?

**Yes; the answer has the correct units of moles $CO_2$.**

Is the number of significant figures correct?

**Yes; the number of significant figures is correct because the mass of $CO_2$ was given to three significant figures.**

Is the answer reasonable?

**Yes; 66 g is about 3/2 the value of the molar mass of $CO_2$. It is reasonable that the sample contains 3/2 (1.5) mol.**

# Practice

**1.** Calculate the number of moles in each of the following masses:

**a.** 3.00 g of boron tribromide, $BBr_3$ **ans: 0.0120 mol $BBr_3$**

**b.** 0.472 g of sodium fluoride, NaF **ans: 0.0112 mol NaF**

**c.** $7.50 \times 10^2$ g of methanol, $CH_3OH$ **ans: 23.4 mol $CH_3OH$**

**d.** 50.0 g of calcium chlorate, $Ca(ClO_3)_2$ **ans: 0.242 mol $Ca(ClO_3)_2$**

**Problem Solving** *continued*

## CONVERTING MOLES OF A COMPOUND TO MASS

Perhaps you have noticed that Sample Problems 1 and 5 are very much alike. In each case, you multiplied the mass by the inverse of the molar mass to calculate the number of moles. The only difference in the two problems is that iron is an element and $CO_2$ is a compound containing a carbon atom and two oxygen atoms.

In Sample Problem 2, you determined the mass of 1.366 mol of zinc. Suppose that you are now asked to determine the mass of 1.366 mol of the molecular compound ammonia, $NH_3$. You can follow the same plan as you did in Sample Problem 2, but this time use the molar mass of ammonia.

## Practice

**1.** Determine the mass of each of the following amounts:

   **a.** 1.366 mol of $NH_3$ **ans: 23.28 g $NH_3$**

   **b.** 0.120 mol of glucose, $C_6H_{12}O_6$ **ans: 21.6 g $C_6H_{12}O_6$**

   **c.** 6.94 mol barium chloride, $BaCl_2$ **ans: 1.45 × 10$^3$ g or 1.45 kg $BaCl_2$**

   **d.** 0.005 mol of propane, $C_3H_8$ **ans: 0.2 g $C_3H_8$**

# Sample Problem 6

Determine the number of molecules in 0.0500 mol of hexane, $C_6H_{14}$.

# Solution

## ANALYZE

What is given in the problem?  **amount of hexane in moles**

What are you asked to find?  **number of molecules of hexane**

| Items | Data |
|-------|------|
| Amount of hexane | 0.0500 mol |
| Avogadro's number—the number of molecules per mole of hexane | $6.022 \times 10^{23}$ molecules/mol |
| Molecules of hexane | ? molecules |

## PLAN

What step is needed to convert from moles of $C_6H_{14}$ to number of molecules of $C_6H_{14}$?

**Avogadro's number is the number of molecules per mole of hexane and can be used to calculate the number of molecules from number of moles.**

$$\boxed{2}\ \text{Amount of } C_6H_{14} \text{ in mol} \xrightarrow[\substack{multiply\ by \\ Avogadro's\ number}]{} \text{Number of } C_6H_{14} \text{ molecules}\ \boxed{3}$$

$$\overset{given}{\text{mol } C_6H_{14}} \times \frac{\overset{Avogadro's\ number}{6.022 \times 10^{23} \text{ molecules } C_6H_{14}}}{1 \text{ mol } C_6H_{14}} = \text{molecules } C_6H_{14}$$

## COMPUTE

$$0.0500 \text{ mol } C_6H_{14} \times \frac{6.022 \times 10^{23} \text{ molecules } C_6H_{14}}{1 \text{ mol } C_6H_{14}} = 3.01 \times 10^{22} \text{ molecules } C_6H_{14}$$

## EVALUATE

Are the units correct?
**Yes; the answer has the correct units of molecules of $C_6H_{14}$.**

Is the number of significant figures correct?
**Yes; three significant figures is correct.**

Is the answer reasonable?
**Yes; multiplying Avogadro's number by 0.05 would yield a product that is a factor of 10 less with a value of $3 \times 10^{22}$.**

| **Problem Solving** *continued*

# Practice

**1.** Calculate the number of molecules in each of the following amounts:

**a.** 4.99 mol of methane, $CH_4$ **ans: $3.00 \times 10^{24}$ molecules $CH_4$**

**b.** 0.005 20 mol of nitrogen gas, $N_2$ **ans: $3.13 \times 10^{21}$ molecules $N_2$**

**c.** 1.05 mol of phosphorus trichloride, $PCl_3$ **ans: $6.32 \times 10^{23}$ molecules $PCl_3$**

**d.** $3.5 \times 10^{-5}$ mol of vitamin C, ascorbic acid, $C_6H_8O_6$ **ans: $2.1 \times 10^{19}$ molecules $C_6H_8O_6$**

**Problem Solving** *continued*

## USING FORMULA UNITS OF IONIC COMPOUNDS

Ionic compounds do not exist as molecules. A crystal of sodium chloride, for example, consists of $Na^+$ ions and $Cl^-$ ions in a 1:1 ratio. Chemists refer to a combination of one $Na^+$ ion and one $Cl^-$ ion as one formula unit of NaCl. A mole of an ionic compound consists of $6.022 \times 10^{23}$ formula units. The mass of one formula unit is called the formula mass. This mass is used in the same way atomic mass or molecular mass is used in calculations.

## Practice

1. Calculate the number of formula units in the following amounts:

   **a.** 1.25 mol of potassium bromide, KBr **ans: $7.53 \times 10^{23}$ formula units KBr**

   **b.** 5.00 mol of magnesium chloride, $MgCl_2$ **ans: $3.01 \times 10^{24}$ formula units $MgCl_2$**

   **c.** 0.025 mol of sodium carbonate, $Na_2CO_3$ **ans: $1.5 \times 10^{22}$ formula units $Na_2CO_3$**

   **d.** $6.82 \times 10^{-6}$ mol of lead(II) nitrate, $Pb(NO_3)_2$ **ans: $4.11 \times 10^{18}$ formula units $Pb(NO_3)_2$**

**| Problem Solving** *continued*

## CONVERTING NUMBER OF MOLECULES OR FORMULA UNITS TO AMOUNT IN MOLES

In Sample Problem 3, you determined the amount in moles of the element lithium. Suppose that you are asked to determine the amount in moles of copper(II) hydroxide in $3.34 \times 10^{34}$ formula units of $Cu(OH)_2$. You can follow the same plan as you did in Sample Problem 3.

# Practice

1. Calculate the amount in moles of the following numbers of molecules or formula units:

   **a.** $3.34 \times 10^{34}$ formula units of $Cu(OH)_2$ **ans: $5.55 \times 10^{10}$ mol $Cu(OH)_2$**

   **b.** $1.17 \times 10^{16}$ molecules of $H_2S$ **ans: $1.94 \times 10^{-8}$ mol $H_2S$**

   **c.** $5.47 \times 10^{21}$ formula units of nickel(II) sulfate, $NiSO_4$ **ans: $9.08 \times 10^{-3}$ mol $NiSO_4$**

   **d.** $7.66 \times 10^{19}$ molecules of hydrogen peroxide, $H_2O_2$ *ans:* **$1.27 \times 10^{-4}$ mol $H_2O_2$**

# Sample Problem 7

**What is the mass of a sample consisting of $1.00 \times 10^{22}$ formula units of $MgSO_4$?**

# Solution

## ANALYZE

What is given in the problem?   **number of magnesium sulfate formula units**

What are you asked to find?   **mass of magnesium sulfate in grams**

| Items | Data |
|---|---|
| Number of formula units of magnesium sulfate | $1.00 \times 10^{22}$ formula units |
| Avogadro's number—the number of formula units of magnesium sulfate per mole | $6.022 \times 10^{23}$ formula units/mol |
| Molar mass of magnesium sulfate | 120.37 g/mol |
| Mass of magnesium sulfate | ? g |

## PLAN

What steps are needed to convert from formula units of $MgSO_4$ to grams of $MgSO_4$?

**First, you must convert the number of formula units of $MgSO_4$ to amount of $MgSO_4$ by using Avogadro's number. Then you can use the molar mass of $MgSO_4$ to convert amount in moles to mass of $MgSO_4$.**

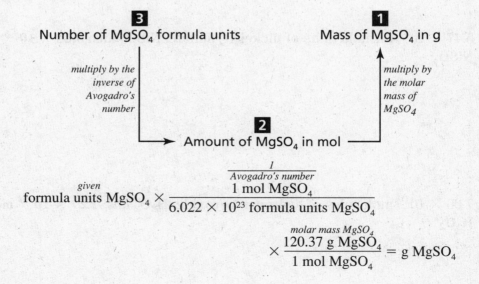

$$\text{formula units } MgSO_4 \times \frac{1 \text{ mol } MgSO_4}{6.022 \times 10^{23} \text{ formula units } MgSO_4}$$

$$\times \frac{120.37 \text{ g } MgSO_4}{1 \text{ mol } MgSO_4} = \text{g } MgSO_4$$

## COMPUTE

$$1.00 \times 10^{22} \text{ formula units MgSO}_4 \times \frac{1 \text{ mol MgSO}_4}{6.022 \times 10^{23} \text{ formula units MgSO}_4}$$

$$\times \frac{120.37 \text{ g MgSO}_4}{1 \text{ mol MgSO}_4} = 2.00 \text{ g MgSO}_4$$

## EVALUATE

Are the units correct?
**Yes; the answer has the correct units of grams of MgSO$_4$.**

Is the number of significant figures correct?
**Yes; the number of significant figures is correct because data were given to three significant figures.**

Is the answer reasonable?
**Yes; 2 g of MgSO$_4$ is about 1/60 of the molar mass of MgSO$_4$. Therefore, 2.00 g MgSO$_4$ will contain about 1/60 of an Avogadro's number of formula units.**

# Practice

1. Calculate the mass of each of the following quantities:

   **a.** $2.41 \times 10^{24}$ molecules of hydrogen, H$_2$ **ans: 8.08 g H$_2$**

   **b.** $5.00 \times 10^{21}$ formula units of aluminum hydroxide, Al(OH)$_3$ **ans: 0.648 g Al(OH)$_3$**

   **c.** $8.25 \times 10^{22}$ molecules of bromine pentafluoride, BrF$_5$ **ans: 24.0 g BrF$_5$**

   **d.** $1.20 \times 10^{23}$ formula units of sodium oxalate, Na$_2$C$_2$O$_4$ **ans: 26.7 g Na$_2$C$_2$O$_4$**

**Problem Solving** *continued*

## CONVERTING MOLECULES OR FORMULA UNITS OF A COMPOUND TO MASS

In Sample Problem 4, you converted a given mass of boron to the number of boron atoms present in the sample. You can now apply the same method to convert mass of an ionic or molecular compound to numbers of molecules or formula units.

## Practice

1. Calculate the number of molecules or formula units in each of the following masses:

   **a.** 22.9 g of sodium sulfide, $Na_2S$ **ans: $1.77 \times 10^{23}$ formula units $Na_2S$**

   **b.** 0.272 g of nickel(II) nitrate, $Ni(NO_3)_2$ **ans: $8.96 \times 10^{20}$ formula units $Ni(NO_3)_2$**

   **c.** 260 mg of acrylonitrile, $CH_2CHCN$ **ans: $3.0 \times 10^{21}$ molecules $CH_2CHCN$**

**Problem Solving** *continued*

# Additional Problems

**1.** Calculate the number of moles in each of the following masses:

**a.** 0.039 g of palladium

**b.** 8200 g of iron

**c.** 0.0073 kg of tantalum

**d.** 0.006 55 g of antimony

**e.** 5.64 kg of barium

**f.** $3.37 \times 10^{-6}$ g of molybdenum

**2.** Calculate the mass in grams of each of the following amounts:

**a.** 1.002 mol of chromium

**b.** 550 mol of aluminum

**c.** $4.08 \times 10^{-8}$ mol of neon

**d.** 7 mol of titanium

**e.** 0.0086 mol of xenon

**f.** $3.29 \times 10^4$ mol of lithium

**3.** Calculate the number of atoms in each of the following amounts:

**a.** 17.0 mol of germanium

**b.** 0.6144 mol of copper

**c.** 3.02 mol of tin

**d.** $2.0 \times 10^6$ mol of carbon

**e.** 0.0019 mol of zirconium

**f.** $3.227 \times 10^{-10}$ mol of potassium

**4.** Calculate the number of moles in each of the following quantities:

**a.** $6.022 \times 10^{24}$ atoms of cobalt

**b.** $1.06 \times 10^{23}$ atoms of tungsten

**c.** $3.008 \times 10^{19}$ atoms of silver

**d.** 950 000 000 atoms of plutonium

**e.** $4.61 \times 10^{17}$ atoms of radon

**f.** 8 trillion atoms of cerium

**5.** Calculate the number of atoms in each of the following masses:

**a.** 0.0082 g of gold

**b.** 812 g of molybdenum

**c.** $2.00 \times 10^2$ mg of americium

**d.** 10.09 kg of neon

**e.** 0.705 mg of bismuth

**f.** 37 μg of uranium

| **Problem Solving** *continued*

**6.** Calculate the mass of each of the following:

   **a.** $8.22 \times 10^{23}$ atoms of rubidium

   **b.** 4.05 Avogadro's numbers of manganese atoms

   **c.** $9.96 \times 10^{26}$ atoms of tellurium

   **d.** 0.000 025 Avogadro's numbers of rhodium atoms

   **e.** 88 300 000 000 000 atoms of radium

   **f.** $2.94 \times 10^{17}$ atoms of hafnium

**7.** Calculate the number of moles in each of the following masses:

   **a.** 45.0 g of acetic acid, $CH_3COOH$

   **b.** 7.04 g of lead(II) nitrate, $Pb(NO_3)_2$

   **c.** 5000 kg of iron(III) oxide, $Fe_2O_3$

   **d.** 12.0 mg of ethylamine, $C_2H_5NH_2$

   **e.** 0.003 22 g of stearic acid, $C_{17}H_{35}COOH$

   **f.** 50.0 kg of ammonium sulfate, $(NH_4)_2SO_4$

**8.** Calculate the mass of each of the following amounts:

   **a.** 3.00 mol of selenium oxybromide, $SeOBr_2$

   **b.** 488 mol of calcium carbonate, $CaCO_3$

   **c.** 0.0091 mol of retinoic acid, $C_{20}H_{28}O_2$

   **d.** $6.00 \times 10^{-8}$ mol of nicotine, $C_{10}H_{14}N_2$

   **e.** 2.50 mol of strontium nitrate, $Sr(NO_3)_2$

   **f.** $3.50 \times 10^{-6}$ mol of uranium hexafluoride, $UF_6$

**9.** Calculate the number of molecules or formula units in each of the following amounts:

   **a.** 4.27 mol of tungsten(VI) oxide, $WO_3$

   **b.** 0.003 00 mol of strontium nitrate, $Sr(NO_3)_2$

   **c.** 72.5 mol of toluene, $C_6H_5CH_3$

   **d.** $5.11 \times 10^{-7}$ mol of α-tocopherol (vitamin E), $C_{29}H_{50}O_2$

   **e.** 1500 mol of hydrazine, $N_2H_4$

   **f.** 0.989 mol of nitrobenzene $C_6H_5NO_2$

**10.** Calculate the number of molecules or formula units in each of the following masses:

   **a.** 285 g of iron(III) phosphate, $FePO_4$

   **b.** 0.0084 g of $C_5H_5N$

   **c.** 85 mg of 2-methyl-1-propanol, $(CH_3)_2CHCH_2OH$

   **d.** $4.6 \times 10^{-4}$ g of mercury(II) acetate, $Hg(C_2H_3O_2)_2$

   **e.** 0.0067 g of lithium carbonate, $Li_2CO_3$

**| Problem Solving** *continued*

11. Calculate the mass of each of the following quantities:

   **a.** $8.39 \times 10^{23}$ molecules of fluorine, $F_2$

   **b.** $6.82 \times 10^{24}$ formula units of beryllium sulfate, $BeSO_4$

   **c.** $7.004 \times 10^{26}$ molecules of chloroform, $CHCl_3$

   **d.** 31 billion formula units of chromium(III) formate, $Cr(CHO_2)_3$

   **e.** $6.3 \times 10^{18}$ molecules of nitric acid, $HNO_3$

   **f.** $8.37 \times 10^{25}$ molecules of freon 114, $C_2Cl_2F_4$

12. Precious metals are commonly measured in troy ounces. A troy ounce is equivalent to 31.1 g. How many moles are in a troy ounce of gold? How many moles are in a troy ounce of platinum? of silver?

13. A chemist needs 22.0 g of phenol, $C_6H_5OH$, for an experiment. How many moles of phenol is this?

14. A student needs 0.015 mol of iodine crystals, $I_2$, for an experiment. What mass of iodine crystals should the student obtain?

15. The weight of a diamond is given in carats. One carat is equivalent to 200. mg. A pure diamond is made up entirely of carbon atoms. How many carbon atoms make up a 1.00 carat diamond?

16. 8.00 g of calcium chloride, $CaCl_2$, is dissolved in 1.000 kg of water.

   **a.** How many moles of $CaCl_2$ are in solution? How many moles of water are present?

   **b.** Assume that the ionic compound, $CaCl_2$, separates completely into $Ca^{2+}$ and $Cl^-$ ions when it dissolves in water. How many moles of each ion are present in the solution?

17. How many moles are in each of the following masses?

   **a.** 453.6 g (1.000 pound) of sucrose (table sugar), $C_{12}H_{22}O_{11}$

   **b.** 1.000 pound of table salt, NaCl

18. When the ionic compound $NH_4Cl$ dissolves in water, it breaks into one ammonium ion, $NH_4^+$, and one chloride ion, $Cl^-$. If you dissolved 10.7 g of $NH_4Cl$ in water, how many moles of ions would be in solution?

19. What is the total amount in moles of atoms in a jar that contains $2.41 \times 10^{24}$ atoms of chromium, $1.51 \times 10^{23}$ atoms of nickel, and $3.01 \times 10^{23}$ atoms of copper?

20. The density of liquid water is 0.997 g/mL at 25°C.

   **a.** Calculate the mass of 250.0 mL (about a cupful) of water.

   **b.** How many moles of water are in 250.0 mL of water? Hint: Use the result of (a).

   **c.** Calculate the volume that would be occupied by 2.000 mol of water at 25°C.

   **d.** What mass of water is 2.000 mol of water?

21. An Avogadro's number (1 mol) of sugar molecules has a mass of 342 g, but an Avogadro's number (1 mol) of water molecules has a mass of only 18 g. Explain why there is such a difference between the mass of 1 mol of sugar and the mass of 1 mol of water.

22. Calculate the mass of aluminum that would have the same number of atoms as 6.35 g of cadmium.

23. A chemist weighs a steel cylinder of compressed oxygen, $O_2$, and finds that it has a mass of 1027.8 g. After some of the oxygen is used in an experiment, the cylinder has a mass of 1023.2 g. How many moles of oxygen gas are used in the experiment?

24. Suppose that you could decompose 0.250 mol of $Ag_2S$ into its elements.

   **a.** How many moles of silver would you have? How many moles of sulfur would you have?

   **b.** How many moles of $Ag_2S$ are there in 38.8 g of $Ag_2S$? How many moles of silver and sulfur would be produced from this amount of $Ag_2S$?

   **c.** Calculate the masses of silver and sulfur produced in (b).

# Problem Solving

## Percentage Composition

Suppose you are working in an industrial laboratory. Your supervisor gives you a bottle containing a white crystalline compound and asks you to determine its identity. Several unlabeled drums of this substance have been discovered in a warehouse, and no one knows what it is. You take it into the laboratory and carry out an analysis, which shows that the compound is composed of the elements sodium, carbon, and oxygen. Immediately, you think of the compound sodium carbonate, $Na_2CO_3$, a very common substance found in most laboratories and used in many industrial processes.

Before you report your conclusion to your boss, you decide to check a reference book to see if there are any other compounds that contain only the elements sodium, carbon, and oxygen. You discover that there is another compound, sodium oxalate, which has the formula $Na_2C_2O_4$. When you read about this compound, you find that it is highly poisonous and can cause serious illness and even death. Mistaking sodium carbonate for sodium oxalate could have very serious consequences. What can you do to determine the identity of your sample? Is it the common industrial substance or the dangerous poison?

Fortunately, you can determine not only *which* elements are in the compound, but also *how much* of each element is present. As you have learned, every compound has a definite composition. Every water molecule is made up of two hydrogen atoms and one oxygen atom, no matter where the water came from. A formula unit of sodium chloride is composed of one sodium atom and one chlorine atom, no matter whether the salt came from a mine or was obtained by evaporating sea water.

Likewise, sodium carbonate always has two sodium atoms, one carbon atom, and three oxygen atoms per formula unit, giving it the formula $Na_2CO_3$; and a formula unit of sodium oxalate always contains two sodium atoms, two carbon atoms, and four oxygen atoms, giving it the formula $Na_2C_2O_4$. Because each atom has a definite mass, each compound will have a distinct composition by mass. This composition is usually expressed as the percentage composition of the compound—the percentage by mass of each element in a compound. To identify a compound, you can compare the percentage composition obtained by laboratory analysis with a calculated percentage composition of each possible compound.

**General Plan for Determining Percentage
Composition of a Compound**

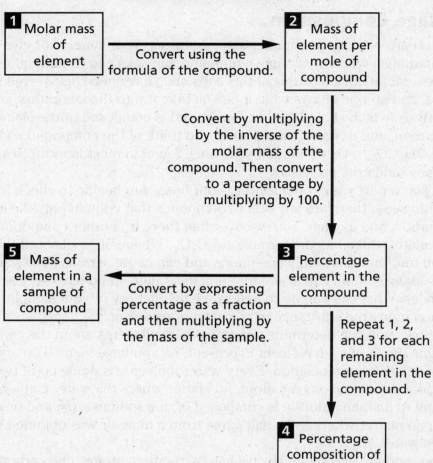

1 Molar mass
of
element

Convert using the
formula of the compound.

2 Mass of
element per
mole of
compound

Convert by multiplying
by the inverse of the
molar mass of the
compound. Then convert
to a percentage by
multiplying by 100.

5 Mass of
element in a
sample of
compound

Convert by expressing
percentage as a fraction
and then multiplying by
the mass of the sample.

3 Percentage
element in the
compound

Repeat 1, 2,
and 3 for each
remaining
element in the
compound.

4 Percentage
composition of
the compound

**| Problem Solving** *continued*

# Sample Problem 1

Determine the percentage composition of sodium carbonate, $Na_2CO_3$.

# Solution

## ANALYZE

What is given in the problem?     **the formula of sodium carbonate**

What are you asked to find?       **the percentage of each element in sodium carbonate (the percentage composition)**

| Items | Data |
|-------|------|
| Formula of sodium carbonate | $Na_2CO_3$ |
| Molar mass of each element* | Na = 22.99 g/mol<br>C = 12.01 g/mol<br>O = 16.00 g/mol |
| Molar mass of sodium carbonate | 105.99 g/mol |
| Percentage composition of sodium carbonate | ?% |

*determined from the periodic table

## PLAN

What step is needed to determine the mass of each element per mole of compound?

**Multiply the molar mass of each element by the ratio of the number of moles of that element in a mole of the compound (the subscript of that element in the compound's formula).**

What steps are needed to determine the portion of each element as a percentage of the mass of the compound?

**Multiply the mass of each element by the inverse of the molar mass of the compound, and then multiply by 100 to convert to a percentage.**

## Step 1

**1** Molar mass of Na $\xrightarrow{\text{\scriptsize multiply by the subscript of Na in } Na_2CO_3}$ **2** Mass Na per mole $Na_2CO_3$

$$\underset{\text{molar mass Na}}{\frac{22.99 \text{ g Na}}{1 \text{ mol Na}}} \times \underset{\substack{\text{ratio of mol Na per mol} \\ Na_2CO_3 \text{ from formula}}}{\frac{2 \text{ mol Na}}{1 \text{ mol } Na_2CO_3}} = \frac{\text{g Na}}{1 \text{ mol } Na_2CO_3}$$

**| Problem Solving** *continued*

## Step 2

**2** Mass Na per mole $Na_2CO_3$ —→ *multiply by the inverse of the molar mass of $Na_2CO_3$ and multiply by 100* —→ **3** Percentage Na in $Na_2CO_3$

$$\underset{\text{from Step 1}}{\frac{\text{g Na}}{1 \text{ mol Na}_2\text{CO}_3}} \times \overset{\frac{1}{\text{molar mass Na}_2\text{CO}_3}}{\frac{1 \text{ mol Na}_2\text{CO}_3}{105.99 \text{ g Na}_2\text{CO}_3}} \times 100 = \text{percentage Na in Na}_2\text{CO}_3$$

Now you can combine Step 1 and Step 2 into one calculation.

**combining Steps 1 and 2**

$$\frac{22.99 \text{ g Na}}{1 \text{ mol Na}} \times \frac{2 \text{ mol Na}}{1 \text{ mol Na}_2\text{CO}_3} \times \frac{1 \text{ mol Na}_2\text{CO}_3}{105.99 \text{ g Na}_2\text{CO}_3} \times 100 = \text{percentage Na in Na}_2\text{CO}_3$$

Finally, determine the percentage of carbon and oxygen in $Na_2CO_3$ by repeating the calculation above with each of those elements.

**3** Percentage of each element in $Na_2CO_3$ —→ *repeat Steps 1 and 2 for each remaining element* —→ **4** Percentage composition

## COMPUTE

**percentage sodium**

$$\frac{22.99 \text{ g Na}}{1 \text{ mol Na}} \times \frac{2 \text{ mol Na}}{1 \text{ mol Na}_2\text{CO}_3} \times \frac{1 \text{ mol Na}_2\text{CO}_3}{105.99 \text{ g Na}_2\text{CO}_3} \times 100 = 43.38\% \text{ Na}$$

**percentage carbon**

$$\frac{12.01 \text{ g C}}{1 \text{ mol C}} \times \frac{1 \text{ mol C}}{1 \text{ mol Na}_2\text{CO}_3} \times \frac{1 \text{ mol Na}_2\text{CO}_3}{105.99 \text{ g Na}_2\text{CO}_3} \times 100 = 11.33\% \text{ C}$$

**percentage oxygen**

$$\frac{16.00 \text{ g O}}{1 \text{ mol O}} \times \frac{3 \text{ mol O}}{1 \text{ mol Na}_2\text{CO}_3} \times \frac{1 \text{ mol Na}_2\text{CO}_3}{105.99 \text{ g Na}_2\text{CO}_3} \times 100 = 45.29\% \text{ O}$$

| Element | Percentage |
|---------|------------|
| sodium  | 43.38% Na  |
| carbon  | 11.33% C   |
| oxygen  | 45.29% O   |

## EVALUATE

Are the units correct?

**Yes; the composition is given in percentages.**

Is the number of significant figures correct?

**Yes; four significant figures is correct because the molar masses have four significant figures.**

Is the answer reasonable?

**Yes; the percentages add up to 100 percent.**

# Practice

**1.** Determine the percentage composition of each of the following compounds:

**a.** sodium oxalate, $Na_2C_2O_4$ **ans: 34.31% Na, 17.93% C, 47.76% O**

**b.** ethanol, $C_2H_5OH$ **ans: 52.13% C, 13.15% H, 34.72% O**

**c.** aluminum oxide, $Al_2O_3$ **ans: 52.92% Al, 47.08% O**

**d.** potassium sulfate, $K_2SO_4$ **ans: 44.87% K, 18.40% S, 36.72% O**

**2.** Suppose that your laboratory analysis of the white powder discussed at the beginning of this chapter showed 42.59% Na, 12.02% C, and 44.99% oxygen. Would you report that the compound is sodium oxalate or sodium carbonate (use the results of Practice Problem 1 and Sample Problem 1)? **ans: sodium carbonate**

| Problem Solving *continued*

## Sample Problem 2

**Calculate the mass of zinc in a 30.00 g sample of zinc nitrate, $Zn(NO_3)_2$.**

## Solution

### ANALYZE

What is given in the problem?　　**the mass in grams of zinc nitrate**

What are you asked to find?　　**the mass in grams of zinc in the sample**

| Items | Data |
|---|---|
| Mass of zinc nitrate | 30.00 g |
| Formula of zinc nitrate | $Zn(NO_3)_2$ |
| Molar mass of zinc nitrate | 189.41 g/mol |
| Mass of zinc in the sample | ? g |

### PLAN

What steps are needed to determine the mass of Zn in a given mass of $Zn(NO_3)_2$?

**The percentage of Zn in $Zn(NO_3)_2$ can be calculated and used to find the mass of Zn in the sample.**

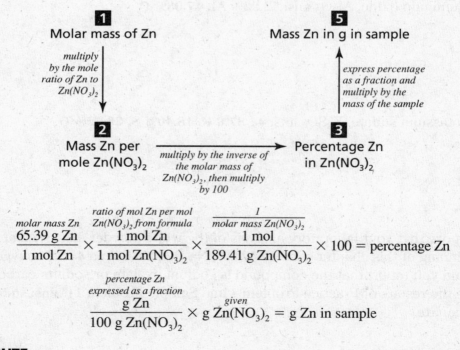

$$\underset{\text{molar mass Zn}}{\frac{65.39 \text{ g Zn}}{1 \text{ mol Zn}}} \times \underset{\substack{\text{ratio of mol Zn per mol} \\ Zn(NO_3)_2 \text{ from formula}}}{\frac{1 \text{ mol Zn}}{1 \text{ mol } Zn(NO_3)_2}} \times \underset{\frac{1}{\text{molar mass } Zn(NO_3)_2}}{\frac{1 \text{ mol}}{189.41 \text{ g } Zn(NO_3)_2}} \times 100 = \text{percentage Zn}$$

$$\underset{\substack{\text{percentage Zn} \\ \text{expressed as a fraction}}}{\frac{\text{g Zn}}{100 \text{ g } Zn(NO_3)_2}} \times \overset{given}{\text{g } Zn(NO_3)_2} = \text{g Zn in sample}$$

### COMPUTE

$$\frac{65.39 \text{ g Zn}}{1 \text{ mol Zn}} \times \frac{1 \text{ mol Zn}}{1 \text{ mol } Zn(NO_3)_2} \times \frac{1 \text{ mol } Zn(NO_3)_2}{189.41 \text{ g } Zn(NO_3)_2} \times 100 = 34.52\% \text{ Zn}$$

Note that mass percentage is the same as grams per 100 g, so 34.52% Zn in $Zn(NO_3)_2$ is the same as 34.52 g Zn in 100 g $Zn(NO_3)_2$.

$$\frac{34.52 \text{ g Zn}}{100 \text{ g } Zn(NO_3)_2} \times 30.00 \text{ g } Zn(NO_3)_2 = 10.36 \text{ g Zn}$$

## EVALUATE
Are the units correct?
**Yes; units cancel to give the correct units, grams of zinc.**

Is the number of significant figures correct?
**Yes; four significant figures is correct because the data given have four significant figures.**

Is the answer reasonable?
**Yes; the molar mass of zinc is about one third of the molar mass of $Zn(NO_3)_2$, and 10.36 g Zn is about one third of 30.00 g of $Zn(NO_3)_2$.**

# Practice

1. Calculate the mass of the given element in each of the following compounds:

   **a.** bromine in 50.0 g potassium bromide, KBr **ans: 33.6 g Br**

   **b.** chromium in 1.00 kg sodium dichromate, $Na_2Cr_2O_7$ **ans: 397 g Cr**

   **c.** nitrogen in 85.0 mg of the amino acid lysine, $C_6H_{14}N_2O_2$ **ans: 16.3 mg N**

   **d.** cobalt in 2.84 g cobalt(II) acetate, $Co(C_2H_3O_2)_2$ **ans: 0.945 g Co**

# HYDRATES

Many compounds, especially ionic compounds, are produced and purified by crystallizing them from water solutions. When this happens, some compounds incorporate water molecules into their crystal structure. These crystalline compounds are called *hydrates* because they include water molecules. The number of water molecules per formula unit is specific for each type of crystal. When you have to measure a certain quantity of the compound, it is important to know how much the water molecules contribute to the mass.

You may have seen blue crystals of copper(II) sulfate in the laboratory. When this compound is crystallized from water solution, the crystals include five water molecules for each formula unit of $CuSO_4$. The true name of the substance is copper(II) sulfate pentahydrate, and its formula is written correctly as $CuSO_4 \cdot 5H_2O$. Notice that the five water molecules are written separately. They are preceded by a dot, which means they are attached to the copper sulfate molecule. On a molar basis, a mole of $CuSO_4 \cdot 5H_2O$ contains 5 mol of water per mole of $CuSO_4 \cdot 5H_2O$. The water molecules contribute to the total mass of $CuSO_4 \cdot 5H_2O$. When you determine the percentage water in a hydrate, the water molecules are treated separately, as if they were another element.

## Sample Problem 3

Determine the percentage water in copper(II) sulfate pentahydrate, $CuSO_4 \cdot 5H_2O$.

## Solution

### ANALYZE

What is given in the problem?     **the formula of copper(II) sulfate pentahydrate**

What are you asked to find?     **the percentage water in the hydrate**

| Items | Data |
|---|---|
| Formula of copper(II) sulfate pentahydrate | $CuSO_4 \cdot 5H_2O$ |
| Molar mass of $H_2O$ | 18.02 g/mol |
| Molar mass of copper(II) sulfate pentahydrate* | 249.72 g/mol |
| Percentage $H_2O$ in $CuSO_4 \cdot 5H_2O$ | ?% |

*molar mass of $CuSO_4$ + mass of 5 mol $H_2O$

**Problem Solving** *continued*

## PLAN

What steps are needed to determine the percentage of water in $CuSO_4 \cdot 5H_2O$?
**Find the mass of water per mole of hydrate, multiply by the inverse molar mass of the hydrate, and multiply that by 100 to convert to a percentage.**

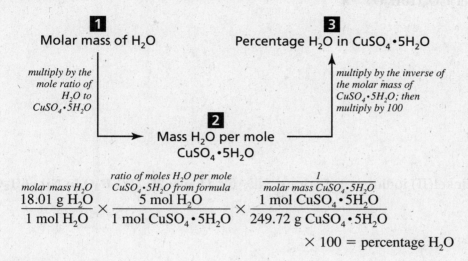

**1** Molar mass of $H_2O$

*multiply by the mole ratio of $H_2O$ to $CuSO_4 \cdot 5H_2O$*

**2** Mass $H_2O$ per mole $CuSO_4 \cdot 5H_2O$

**3** Percentage $H_2O$ in $CuSO_4 \cdot 5H_2O$

*multiply by the inverse of the molar mass of $CuSO_4 \cdot 5H_2O$; then multiply by 100*

$$\underset{\text{molar mass } H_2O}{\frac{18.01 \text{ g } H_2O}{1 \text{ mol } H_2O}} \times \underset{\substack{\text{ratio of moles } H_2O \text{ per mole} \\ CuSO_4 \cdot 5H_2O \text{ from formula}}}{\frac{5 \text{ mol } H_2O}{1 \text{ mol } CuSO_4 \cdot 5H_2O}} \times \underset{\text{molar mass } CuSO_4 \cdot 5H_2O}{\frac{1 \text{ mol } CuSO_4 \cdot 5H_2O}{249.72 \text{ g } CuSO_4 \cdot 5H_2O}}$$
$$\times \, 100 = \text{percentage } H_2O$$

## COMPUTE

$$\frac{18.01 \text{ g } H_2O}{1 \text{ mol } H_2O} \times \frac{5 \text{ mol } H_2O}{1 \text{ mol } CuSO_4 \cdot 5H_2O} \times \frac{1 \text{ mol } CuSO_4 \cdot 5H_2O}{249.72 \text{ g } CuSO_4 \cdot 5H_2O} \times 100 = 36.08\% \, H_2O$$

## EVALUATE

Are the units correct?
**Yes; the percentage of water in copper(II) sulfate pentahydrate was needed.**

Is the number of significant figures correct?
**Yes; four significant figures is correct because molar masses were given to at least four significant figures.**

Is the answer reasonable?
**Yes; five water molecules have a mass of about 90 g, and 90 g is a little more than 1/3 of 250 g; the calculated percentage is a little more than 1/3.**

**❚ Problem Solving** *continued*

## Practice

**1.** Calculate the percentage of water in each of the following hydrates:

    **a.** sodium carbonate decahydrate, $Na_2CO_3 \cdot 10H_2O$ **ans: 62.97% $H_2O$ in $Na_2CO_3 \cdot 10H_2O$**

    **b.** nickel(II) iodide hexahydrate, $NiI_2 \cdot 6H_2O$ **ans: 25.71% $H_2O$ in $NiI_2 \cdot 6H_2O$**

    **c.** ammonium hexacyanoferrate(III) trihydrate (commonly called ammonium ferricyanide), $(NH_4)_2Fe(CN)_6 \cdot 3H_2O$ **ans: 17.89 % $H_2O$ in $(NH_4)_2Fe(CN)_6 \cdot 3H_2O$**

    **d.** aluminum bromide hexahydrate **ans: 28.85% $H_2O$ in $AlBr_3 \cdot 6H_2O$**

# Additional Problems

1. Write formulas for the following compounds and determine the percentage composition of each:
   a. nitric acid
   b. ammonia
   c. mercury(II) sulfate
   d. antimony(V) fluoride

2. Calculate the percentage composition of the following compounds:
   a. lithium bromide, LiBr
   b. anthracene, $C_{14}H_{10}$
   c. ammonium nitrate, $NH_4NO_3$
   d. nitrous acid, $HNO_2$
   e. silver sulfide, $Ag_2S$
   f. iron(II) thiocyanate, $Fe(SCN)_2$
   g. lithium acetate
   h. nickel(II) formate

3. Calculate the percentage of the given element in each of the following compounds:
   a. nitrogen in urea, $NH_2CONH_2$
   b. sulfur in sulfuryl chloride, $SO_2Cl_2$
   c. thallium in thallium(III) oxide, $Tl_2O_3$
   d. oxygen in potassium chlorate, $KClO_3$
   e. bromine in calcium bromide, $CaBr_2$
   f. tin in tin(IV) oxide, $SnO_2$

4. Calculate the mass of the given element in each of the following quantities:
   a. oxygen in 4.00 g of manganese dioxide, $MnO_2$
   b. aluminum in 50.0 metric tons of aluminum oxide, $Al_2O_3$
   c. silver in 325 g silver cyanide, AgCN
   d. gold in 0.780 g of gold(III) selenide, $Au_2Se_3$
   e. selenium in 683 g sodium selenite, $Na_2SeO_3$
   f. chlorine in $5.0 \times 10^4$ g of 1,1-dichloropropane, $CHCl_2CH_2CH_3$

5. Calculate the percentage of water in each of the following hydrates:
   a. strontium chloride hexahydrate, $SrCl_2 \cdot 6H_2O$
   b. zinc sulfate heptahydrate, $ZnSO_4 \cdot 7H_2O$
   c. calcium fluorophosphate dihydrate, $CaFPO_3 \cdot 2H_2O$
   d. beryllium nitrate trihydrate, $Be(NO_3)_2 \cdot 3H_2O$

**6.** Calculate the percentage of the given element in each of the following hydrates. You must first determine the formulas of the hydrates.

   **a.** nickel in nickel(II) acetate tetrahydrate

   **b.** chromium in sodium chromate tetrahydrate

   **c.** cerium in cerium(IV) sulfate tetrahydrate

**7.** Cinnabar is a mineral that is mined in order to produce mercury. Cinnabar is mercury(II) sulfide, $HgS$. What mass of mercury can be obtained from 50.0 kg of cinnabar?

**8.** The minerals malachite, $Cu_2(OH)_2CO_3$, and chalcopyrite, $CuFeS_2$, can be mined to obtain copper metal. How much copper could be obtained from $1.00 \times 10^3$ kg of each? Which of the two has the greater copper content?

**9.** Calculate the percentage of the given element in each of the following hydrates:

   **a.** vanadium in vanadium oxysulfate dihydrate, $VOSO_4 \cdot 2H_2O$

   **b.** tin in potassium stannate trihydrate, $K_2SnO_3 \cdot 3H_2O$

   **c.** chlorine in calcium chlorate dihydrate, $CaClO_3 \cdot 2H_2O$

**10.** Heating copper sulfate pentahydrate will evaporate the water from the crystals, leaving anhydrous copper sulfate, a white powder. *Anhydrous* means "without water." What mass of anhydrous $CuSO_4$ would be produced by heating 500.0 g of $CuSO_4 \cdot 5H_2O$?

**11.** Silver metal may be precipitated from a solution of silver nitrate by placing a copper strip into the solution. What mass of $AgNO_3$ would you dissolve in water in order to get 1.00 g of silver?

**12.** A sample of $Ag_2S$ has a mass of 62.4 g. What mass of each element could be obtained by decomposing this sample?

**13.** A quantity of epsom salts, magnesium sulfate heptahydrate, $MgSO_4 \cdot 7H_2O$, is heated until all the water is driven off. The sample loses 11.8 g in the process. What was the mass of the original sample?

**14.** The process of manufacturing sulfuric acid begins with the burning of sulfur. What mass of sulfur would have to be burned in order to produce 1.00 kg of $H_2SO_4$? Assume that all of the sulfur ends up in the sulfuric acid.

# Problem Solving

## Empirical Formulas

Suppose you analyze an unknown compound that is a white powder and find that it is composed of 36.5% sodium, 38.1% oxygen, and 25.4% sulfur. You can use those percentages to determine the mole ratios among sodium, sulfur, and oxygen and write a formula for the compound.

To begin, the mass percentages of each element can be interpreted as "grams of element per 100 grams of compound." To make things simpler, you can assume you have a 100 g sample of the unknown compound. The unknown compound contains 36.5% sodium by mass. Therefore 100.0 g of the compound would contain 36.5 g of sodium. You already know how to convert mass of a substance into number of moles, so you can calculate the number of moles of sodium in 36.5 g. After you find the number of moles of each element, you can look for a simple ratio among the elements and use this ratio of elements to write a formula for the compound.

The chemical formula obtained from the mass percentages is in the simplest form for that compound. The mole ratios for each element, which you determined from the analytical data given, are reduced to the smallest whole numbers. This simplest formula is also called the empirical formula. The actual formula for the compound could be a multiple of the empirical formula. For instance, suppose you analyze a compound and find that it is composed of 40.0% carbon, 6.7% hydrogen, and 53.3% oxygen. If you determine the formula for this compound based only on the analytical data, you will determine the formula to be $CH_2O$. There are, however, other possibilities for the formula. It could be $C_2H_4O_2$ and still have the same percentage composition. In fact, it could be any multiple of $CH_2O$.

It is possible to convert from the empirical formula to the actual chemical formula for the compound as long as the molar mass of the compound is known. Look again at the $CH_2O$ example. If the true compound were $CH_2O$, it would have a molar mass of 30.03 g/mol. If you do more tests on the unknown compound and find that its molar mass is 60.06, you know that $CH_2O$ cannot be its true identity. The molar mass 60.06 is twice the molar mass of $CH_2O$. Therefore, you know that the true chemical formula must be twice the empirical formula, $(CH_2O) \times 2$, or $C_2H_4O_2$. Any correct molecular formula can be determined from an empirical formula and a molar mass in this same way.

### General Plan for Determining Empirical Formulas and Molecular Formulas

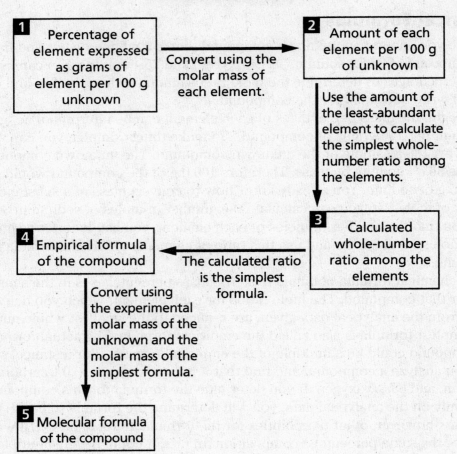

# Sample Problem 1

Determine the empirical formula for an unknown compound composed of 36.5% sodium, 38.1% oxygen, and 25.4% sulfur by mass.

# Solution

## ANALYZE

What is given in the problem?      **the percentage composition of the compound**

What are you asked to find?      **the empirical formula for the compound**

| Items | Data |
|---|---|
| The percentage composition of the unknown subtance | 36.5% sodium<br>38.1% oxygen<br>25.4% sulfur |
| The molar mass of each element* | 22.99 g Na/mol Na<br>16.00 g O/mol O<br>32.07 g S/mol S |
| Amount of each element per 100.0 g of the unknown | ? mol |
| Simplest mole ratio of elements in the unknown | ? |

\* determined from the periodic table

## PLAN

What steps are needed to calculate the amount in moles of each element per 100.0 g of unknown?
**State the percentage of the element in grams and multiply by the inverse of the molar mass of the element.**

What steps are needed to determine the whole-number mole ratio of the elements in the unknown (the simplest formula)?
**Divide the amount of each element by the amount of the least-abundant element. If necessary, multiply the ratio by a small integer that will produce a whole-number ratio.**

<table>
<tr><td>**1**</td><td></td><td>**2**</td></tr>
<tr>
<td>Mass of Na per<br>100.0 g unknown</td>
<td>→<br>*multiply by the inverse of<br>the molar mass of Na*</td>
<td>Amount Na in mol per<br>100.0 g unknown</td>
</tr>
</table>

$$\underset{\substack{percent\ of\ Na\ stated\ as\ grams \\ Na\ per\ 100\ g\ unknown}}{\frac{36.5\text{ g Na}}{100.0\text{ g unknown}}} \times \underset{\substack{1 \\ molar\ mass\ Na}}{\frac{1\text{ mol Na}}{22.99\text{ g Na}}} = \frac{\text{mol Na}}{100.0\text{ g unknown}}$$

Repeat this step for the remaining elements.

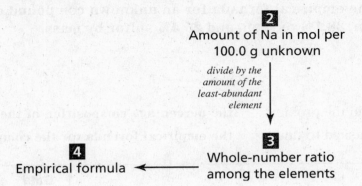

**2**
Amount of Na in mol per
100.0 g unknown

*divide by the
amount of the
least-abundant
element*

**4**
Empirical formula ⟵

**3**
Whole-number ratio
among the elements

## COMPUTE

$$\frac{36.5\ \text{g Na}}{100.0\ \text{g unknown}} \times \frac{1\ \text{mol Na}}{22.99\ \text{g Na}} = \frac{1.59\ \text{mol Na}}{100.0\ \text{g unknown}}$$

$$\frac{38.1\ \text{g O}}{100.0\ \text{g unknown}} \times \frac{1\ \text{mol O}}{16.00\ \text{g O}} = \frac{2.38\ \text{mol O}}{100.0\ \text{g unknown}}$$

$$\frac{25.4\ \text{g S}}{100.0\ \text{g unknown}} \times \frac{1\ \text{mol S}}{32.07\ \text{g S}} = \frac{0.792\ \text{mol S}}{100.0\ \text{g unknown}}$$

Divide the amount of each element by the amount of the least-abundant element, which in this example is S. This can be accomplished by multiplying the amount of each element by the inverse of the amount of the least abundant element.

$$\frac{1.59\ \text{mol Na}}{100.0\ \text{g unknown}} \times \frac{100.0\ \text{g unknown}}{0.792\ \text{mol S}} = \frac{2.01\ \text{mol Na}}{1\ \text{mol S}}$$

$$\frac{2.38\ \text{mol O}}{100.0\ \text{g unknown}} \times \frac{100.0\ \text{g unknown}}{0.792\ \text{mol S}} = \frac{3.01\ \text{mol O}}{1\ \text{mol S}}$$

$$\frac{0.792\ \text{mol S}}{100.0\ \text{g unknown}} \times \frac{100.0\ \text{g unknown}}{0.792\ \text{mol S}} = \frac{1.00\ \text{mol S}}{1\ \text{mol S}}$$

From the calculations, the simplest mole ratio is 2 mol Na:3 mol O:1 mol S.
The simplest formula is therefore $Na_2O_3S$. Seeing the ratio 3 mol O:1 mol S, you can use your knowledge of chemistry to suggest that this possibly represents a sulfite group, $-SO_3$ and propose the formula $Na_2SO_3$.

## EVALUATE

Are the units correct?
**Yes; units canceled throughout the calculation, so it is reasonable to assume that the resulting ratio is accurate.**

Is the number of significant figures correct?
**Yes; ratios were calculated to three significant figures because percentages were given to three significant figures.**

**Problem Solving** *continued*

Is the answer reasonable?
**Yes; the formula, Na$_2$SO$_3$ is plausible, given the mole ratios and considering that the sulfite ion has a 2− charge and the sodium ion has a 1+ charge.**

# Practice

**1.** Determine the empirical formula for compounds that have the following analyses:

   **a.** 28.4% copper, 71.6% bromine **ans: CuBr$_2$**

   **b.** 39.0% potassium, 12.0% carbon, 1.01% hydrogen, and 47.9% oxygen

       **ans: KHCO$_3$**

   **c.** 77.3% silver, 7.4% phosphorus, 15.3% oxygen **ans: Ag$_3$PO$_4$**

   **d.** 0.57% hydrogen, 72.1% iodine, 27.3% oxygen **ans: HIO$_3$**

## Sample Problem 2

**Determine the empirical formula for an unknown compound composed of 38.4% oxygen, 23.7% carbon, and 1.66% hydrogen.**

## Solution

### ANALYZE

What is given in the problem?     **the percentage composition of the compound**

What are you asked to find?       **the empirical formula for the compound**

### PLAN

What steps are needed to calculate the amount in moles of each element per 100.0 g of unknown?

**State the percentage of the element in grams and multiply by the inverse of the molar mass of the element.**

What steps are needed to determine the whole-number mole ratio of the elements in the unknown (the simplest formula)?

**Divide the amount of each element by the amount of the least-abundant element. If necessary, multiply the ratio by a small integer to produce a whole-number ratio.**

<br/>

| **1** |  | **2** |
|---|---|---|
| Mass of K in g per 100.0 g unknown | *multiply by the inverse of the molar mass of K* → | Amount of K in mol per 100.0 g unknown |

*divide by the amount of the least-abundant element, and multiply by an integer that will produce a whole-number ratio*

| **4** |  | **3** |
|---|---|---|
| Empirical formula | ← | Whole-number ratio among the elements |

### COMPUTE

$$\frac{38.4 \text{ g K}}{100.0 \text{ g unknown}} \times \frac{1 \text{ mol K}}{39.10 \text{ g K}} = \frac{0.982 \text{ mol K}}{100.0 \text{ g unknown}}$$

Proceed to find the amount in moles per 100.0 g of unknown for the elements carbon, oxygen, and hydrogen, as in Sample Problem 1.

When determining the formula of a compound having more than two elements, it is usually advisable to put the data and results in a table.

**Problem Solving** *continued*

| Element | Mass per 100.0 g of unknown | Molar mass | Amount in mol per 100.0 g of unknown |
|---------|------------------------------|------------|--------------------------------------|
| Potassium | 38.4 g K | 39.10 g/mol | 0.982 mol K |
| Carbon | 23.7 g C | 12.01 g/mol | 1.97 mol C |
| Oxygen | 36.3 g O | 16.00 g/mol | 2.27 mol O |
| Hydrogen | 1.66 g H | 1.01 g/mol | 1.64 mol H |

Again, as in Sample Problem 1, divide each result by the amount in moles of the least-abundant element, which in this example is K.

You should get the following results:

| Element | Amount in mol of element per 100.0 g of unknown | Amount in mol of element per mol of potassium |
|---------|------------------------------------------------|-----------------------------------------------|
| Potassium | 0.982 mol K | 1.00 mol K |
| Carbon | 1.97 mol C | 2.01 mol C |
| Oxygen | 2.27 mol O | 2.31 mol O |
| Hydrogen | 1.64 mol H | 1.67 mol H |

In contrast to Sample Problem 1, this calculation does not give a simple whole-number ratio among the elements. To solve this problem, multiply by a small integer that will result in a whole-number ratio. You can pick an integer that you think might work, or you can convert the number of moles to an equivalent fractional number. At this point, you should keep in mind that analytical data is never perfect, so change the number of moles to the fraction that is closest to the decimal number. Then, choose the appropriate integer factor to use. In this case, the fractions are in thirds so a factor of 3 will change the fractions into whole numbers.

| Amount in mol of element per mole of potassium | Fraction nearest the decimal value | Integer factor | Whole-number mole ratio |
|------------------------------------------------|-------------------------------------|----------------|-------------------------|
| 1.00 mol K | 1 mol K | ×3 | 3 mol K |
| 2.01 mol C | 2 mol C | ×3 | 6 mol C |
| 2.31 mol O | 2 1/3 mol O | ×3 | 7 mol O |
| 1.67 mol H | 1 2/3 mol H | ×3 | 5 mol H |

Thus, the simplest formula for the compound is $K_3C_6H_5O_7$, which happens to be the formula for potassium citrate.

**Problem Solving** *continued*

**EVALUATE**

Is the answer reasonable?

**Yes; the formula, $K_3C_6H_5O_7$ is plausible, considering that the potassium ion has a 1+ charge and the citrate polyatomic ion has a 3− charge.**

# Practice

1. Determine the simplest formula for compounds that have the following analyses. The data may not be exact.

   **a.** 36.2% aluminum and 63.8% sulfur **ans: $Al_2S_3$**

   **b.** 93.5% niobium and 6.50% oxygen **ans: $Nb_5O_2$**

   **c.** 57.6% strontium, 13.8% phosphorus, and 28.6% oxygen **ans: $Sr_3P_2O_8$ or $Sr_3(PO_4)_2$**

   **d.** 28.5% iron, 48.6% oxygen, and 22.9% sulfur **ans: $Fe_2S_3O_{12}$ or $Fe_2(SO_4)_3$**

**Problem Solving** *continued*

## Sample Problem 3

A compound is analyzed and found to have the empirical formula $CH_2O$. The molar mass of the compound is found to be 153 g/mol. What is the compound's molecular formula?

## Solution
### ANALYZE

What is given in the problem?  **the empirical formula, and the experimental molar mass**

What are you asked to find?  **the molecular formula of the compound**

| Items | Data |
|---|---|
| Empirical formula of unknown | $CH_2O$ |
| Experimental molar mass of unknown | 153 g/mol |
| Molar mass of empirical formula | 30.03 g/mol |
| Molecular formula of the compound | ? |

### PLAN

What steps are needed to determine the molecular formula of the unknown compound?

**Multiply the experimental molar mass by the inverse of the molar mass of the empirical formula. The subscripts of the empirical formula are multiplied by the whole-number factor obtained.**

$$\frac{153 \text{ g}}{1 \text{ mol unknown}} \times \frac{1 \text{ mol } CH_2O}{30.03 \text{ g}} = \frac{\text{mol } CH_2O}{1 \text{ mol unknown}}$$

### COMPUTE

$$\frac{153 \text{ g}}{1 \text{ mol unknown}} \times \frac{1 \text{ mol } CH_2O}{30.03 \text{ g}} = \frac{5.09 \text{ mol } CH_2O}{1 \text{ mol unknown}}$$

Allowing for a little experimental error, the molecular formula must be five times the empirical formula.

$$\text{Molecular formula} = (CH_2O) \times 5 = C_5H_{10}O_5$$

| **Problem Solving** *continued* |

## EVALUATE

Is the answer reasonable?

**Yes; the calculated molar mass of $C_5H_{10}O_5$ is 150.15, which is close to the experimental molar mass of the unknown. Reference books show that there are several different compounds with the formula $C_5H_{10}O_5$.**

# Practice

1. Determine the molecular formula of each of the following unknown substances:

   **a.** empirical formula $CH_2$, experimental molar mass 28 g/mol **ans: $C_2H_4$**

   **b.** empirical formula $B_2H_5$, experimental molar mass 54 g/mol **ans: $B_4H_{10}$**

   **c.** empirical formula $C_2HCl$, experimental molar mass 179 g/mol **ans: $C_6H_3Cl_3$**

   **d.** empirical formula $C_6H_8O$, experimental molar mass 290 g/mol **ans: $C_{18}H_{24}O_3$**

   **e.** empirical formula $C_3H_2O$, experimental molar mass 216 g/mol **ans: $C_{12}H_8O_4$**

| Problem Solving *continued*

## Additional Problems

1. Determine the empirical formula for compounds that have the following analyses:

   **a.** 66.0% barium and 34.0% chlorine

   **b.** 80.38% bismuth, 18.46% oxygen, and 1.16% hydrogen

   **c.** 12.67% aluminum, 19.73% nitrogen, and 67.60% oxygen

   **d.** 35.64% zinc, 26.18% carbon, 34.88% oxygen, and 3.30% hydrogen

   **e.** 2.8% hydrogen, 9.8% nitrogen, 20.5% nickel, 44.5% oxygen, and 22.4% sulfur

   **f.** 8.09% carbon, 0.34% hydrogen, 10.78% oxygen, and 80.78% bromine

2. Sometimes, instead of percentage composition, you will have the composition of a sample by mass. Use the same method shown in Sample Problem 1, but use the actual mass of the sample instead of assuming a 100 g sample. Determine the empirical formula for compounds that have the following analyses:

   **a.** a 0.858 g sample of an unknown substance is composed of 0.537 g of copper and 0.321 g of fluorine

   **b.** a 13.07 g sample of an unknown substance is composed of 9.48 g of barium, 1.66 g of carbon, and 1.93 g of nitrogen

   **c.** a 0.025 g sample of an unknown substance is composed of 0.0091 g manganese, 0.0106 g oxygen, and 0.0053 g sulfur

3. Determine the empirical formula for compounds that have the following analyses:

   **a.** a 0.0082 g sample contains 0.0015 g of nickel and 0.0067 g of iodine

   **b.** a 0.470 g sample contains 0.144 g of manganese, 0.074 g of nitrogen, and 0.252 g of oxygen

   **c.** a 3.880 g sample contains 0.691 g of magnesium, 1.824 g of sulfur, and 1.365 g of oxygen

   **d.** a 46.25 g sample contains 14.77 g of potassium, 9.06 g of oxygen, and 22.42 g of tin

4. Determine the empirical formula for compounds that have the following analyses:

   **a.** 60.9% As and 39.1% S

   **b.** 76.89% Re and 23.12% O

   **c.** 5.04% H, 35.00% N, and 59.96% O

   **d.** 24.3% Fe, 33.9% Cr, and 41.8% O

   **e.** 54.03% C, 37.81% N, and 8.16% H

   **f.** 55.81% C, 3.90% H, 29.43% F, and 10.85% N

# Problem Solving *continued*

**5.** Determine the molecular formulas for compounds having the following empirical formulas and molar masses:

**a.** $C_2H_4S$; experimental molar mass 179

**b.** $C_2H_4O$; experimental molar mass 176

**c.** $C_2H_3O_2$; experimental molar mass 119

**d.** $C_2H_2O$, experimental molar mass 254

**6.** Use the experimental molar mass to determine the molecular formula for compounds having the following analyses:

**a.** 41.39% carbon, 3.47% hydrogen, and 55.14% oxygen; experimental molar mass 116.07

**b.** 54.53% carbon, 9.15% hydrogen, and 36.32% oxygen; experimental molar mass 88

**c.** 64.27% carbon, 7.19% hydrogen, and 28.54% oxygen; experimental molar mass 168.19

**7.** A 0.400 g sample of a white powder contains 0.141 g of potassium, 0.115 g of sulfur, and 0.144 g of oxygen. What is the empirical formula for the compound?

**8.** A 10.64 g sample of a lead compound is analyzed and found to be made up of 9.65 g of lead and 0.99 g of oxygen. Determine the empirical formula for this compound.

**9.** A 2.65 g sample of a salmon-colored powder contains 0.70 g of chromium, 0.65 g of sulfur, and 1.30 g of oxygen. The molar mass is 392.2. What is the formula of the compound?

**10.** Ninhydrin is a compound that reacts with amino acids and proteins to produce a dark-colored complex. It is used by forensic chemists and detectives to see fingerprints that might otherwise be invisible. Ninhydrin's composition is 60.68% carbon, 3.40% hydrogen, and 35.92% oxygen. What is the empirical formula for ninhydrin?

**11.** Histamine is a substance that is released by cells in response to injury, infection, stings, and materials that cause allergic responses, such as pollen. Histamine causes dilation of blood vessels and swelling due to accumulation of fluid in the tissues. People sometimes take *anti*histamine drugs to counteract the effects of histamine. A sample of histamine having a mass of 385 mg is composed of 208 mg of carbon, 31 mg of hydrogen, and 146 mg of nitrogen. The molar mass of histamine is 111 g/mol. What is the molecular formula for histamine?

**12.** You analyze two substances in the laboratory and discover that each has the empirical formula $CH_2O$. You can easily see that they are different substances because one is a liquid with a sharp, biting odor and the other is an odorless, crystalline solid. How can you account for the fact that both have the same empirical formula?

# Problem Solving

## Stoichiometry

So far in your chemistry course, you have learned that chemists count quantities of elements and compounds in terms of moles and that they relate moles of a substance to mass by using the molar mass. In addition, you have learned to write chemical equations so that they represent the rearrangements of atoms that take place during chemical reactions, and you have learned to balance these equations. In this chapter you will be able to put these separate skills together to accomplish one of the most important tasks of chemistry—using chemical equations to make predictions about the quantities of substances that react or are given off as products and relating those quantities to one another. This process of relating quantities of reactants and products in a chemical reaction to one another is called *stoichiometry*.

First, look at an analogy.

Suppose you need to make several sandwiches to take on a picnic with friends. You decide to make turkey-and-cheese sandwiches using the following "equation:"

2 bread slices + 2 turkey slices + 1 lettuce leaf + 1 cheese slice
$$\rightarrow 1 \text{ turkey-and-cheese sandwich}$$

This equation shows that you need those ingredients in a ratio of $2:2:1:1$, respectively. You can use this equation to predict that you would need 30 turkey slices to make 15 sandwiches or 6 cheese slices to go with 12 turkey slices.

Zinc reacts with oxygen according to the following balanced chemical equation:

$$2Zn + O_2 \rightarrow 2ZnO$$

Like the sandwich recipe, this equation can be viewed as a "recipe" for zinc oxide. It tells you that reacting two zinc atoms with a molecule of oxygen will produce two formula units of zinc oxide. Can you predict how many zinc oxide units could be formed from 500 zinc atoms? Could you determine how many moles of oxygen molecules it would take to react with 4 mol of zinc atoms? What if you had 22 g of zinc and wanted to know how many grams of ZnO could be made from it? Keep in mind that the chemical equation relates amounts, not masses, of products and reactants. The problems in this chapter will show you how to solve problems of this kind.

**Problem Solving** *continued*

## General Plan for Solving Stoichiometry Problems

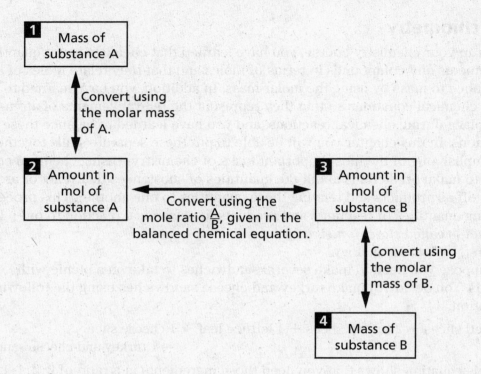

**▌Problem Solving** *continued*

# Sample Problem 1

Ammonia is made industrially by reacting nitrogen and hydrogen under pressure, at high temperature, and in the presence of a catalyst. The equation is $N_2(g) + 3H_2(g) \rightarrow 2NH_3(g)$. If 4.0 mol of $H_2$ react, how many moles of $NH_3$ will be produced?

# Solution

## ANALYZE

What is given in the problem?    **the balanced equation, and the amount of $H_2$ in moles**

What are you asked to find?    **the amount of $NH_3$ produced in moles**

Organization of data is extremely important in dealing with stoichiometry problems. You will find that it is most helpful to make data tables such as the following one.

| Items | Data | |
|---|---|---|
| Substance | $H_2$ | $NH_3$ |
| Coefficient in balanced equation | 3 | 2 |
| Molar mass | NA* | NA |
| Amount | 4.0 mol | ? mol |
| Mass of substance | NA | NA |

*NA means *not applicable to the problem*

## PLAN

What steps are needed to calculate the amount of $NH_3$ that can be produced from 4.0 mol $H_2$?

**Multiply by the mole ratio of $NH_3$ to $H_2$ determined from the coefficients of the balanced equation.**

$$\boxed{2}$$
Amount of $H_2$ in mol $\xrightarrow{\textit{multiply by mole ratio: } \frac{NH_3}{H_2}}$ $\boxed{3}$ Amount of $NH_3$ in mol

$$\overset{\textit{given}}{\text{mol } H_2} \times \frac{\overset{\textit{mole ratio}}{2 \text{ mol } NH_3}}{3 \text{ mol } H_2} = \text{mol } NH_3$$

**▌Problem Solving** *continued*

## COMPUTE

$$4.0 \text{ mol } H_2 \times \frac{2 \text{ mol } NH_3}{3 \text{ mol } H_2} = 2.7 \text{ mol } NH_3$$

## EVALUATE

Are the units correct?

**Yes; the answer has the correct units of moles $NH_3$.**

Is the number of significant figures correct?

**Yes; two significant figures is correct because data were given to two significant figures.**

Is the answer reasonable?

**Yes; the answer is 2/3 of 4.0.**

# Practice

1. How many moles of sodium will react with water to produce 4.0 mol of hydrogen in the following reaction?

$$2Na(s) + 2H_2O(l) \rightarrow 2NaOH(aq) + H_2(g) \text{ ans: 8.0 mol Na}$$

2. How many moles of lithium chloride will be formed by the reaction of chlorine with 0.046 mol of lithium bromide in the following reaction?

$$2LiBr(aq) + Cl_2(g) \rightarrow 2LiCl(aq) + Br_2(l) \text{ ans: 0.046 mol LiCl}$$

**Problem Solving** *continued*

**3.** Aluminum will react with sulfuric acid in the following reaction.

$$2Al(s) + 3H_2SO_4(l) \rightarrow Al_2(SO_4)_3(aq) + 3H_2(g)$$

**a.** How many moles of $H_2SO_4$ will react with 18 mol Al? **ans: 27 mol $H_2SO_4$**

**b.** How many moles of each product will be produced? **ans: 27 mol $H_2$, 9 mol $Al_2(SO_4)_3$**

**4.** Propane burns in excess oxygen according to the following reaction.

$$C_3H_8 + 5O_2 \rightarrow 3CO_2 + 4H_2O$$

**a.** How many moles each of $CO_2$ and $H_2O$ are formed from 3.85 mol of propane? **ans: 11.6 mol $CO_2$, 15.4 mol $H_2O$**

**b.** If 0.647 mol of oxygen is used in the burning of propane, how many moles each of $CO_2$ and $H_2O$ are produced? How many moles of $C_3H_8$ are consumed? **ans: 0.388 mol $CO_2$, 0.518 mol $H_2O$, 0.129 mol $C_3H_8$**

## Sample Problem 2

**Potassium chlorate is sometimes decomposed in the laboratory to generate oxygen. The reaction is $2KClO_3(s) \rightarrow 2KCl(s) + 3O_2(g)$. What mass of $KClO_3$ do you need to produce 0.50 mol $O_2$?**

## Solution

### ANALYZE

What is given in the problem?     **the amount of oxygen in moles**

What are you asked to find?     **the mass of potassium chlorate**

| Items | Data | |
|---|---|---|
| Substance | $KClO_3$ | $O_2$ |
| Coefficient in balanced equation | 2 | 3 |
| Molar mass* | 122.55 g/mol | NA |
| Amount | ? mol | 0.50 mol |
| Mass | ? g | NA |

\* determined from the periodic table

### PLAN

What steps are needed to calculate the mass of $KClO_3$ needed to produce 0.50 mol $O_2$?

**Use the mole ratio to convert amount of $O_2$ to amount of $KClO_3$. Then convert amount of $KClO_3$ to mass of $KClO_3$.**

$$\underset{given}{mol\ O_2} \times \underset{mole\ ratio}{\frac{2\ mol\ KClO_3}{3\ mol\ O_2}} \times \underset{molar\ mass\ KClO_3}{\frac{122.55\ g\ KClO_3}{1\ mol\ KClO_3}} = g\ KClO_3$$

### COMPUTE

$$0.50\ mol\ O_2 \times \frac{2\ mol\ KClO_3}{3\ mol\ O_2} \times \frac{122.55\ g\ KClO_3}{1\ mol\ KClO_3} = 41\ g\ KClO_3$$

### EVALUATE

Are the units correct?

**Yes; units canceled to give grams of $KClO_3$.**

Is the number of significant figures correct?
**Yes; two significant figures is correct.**

Is the answer reasonable?
**Yes; 41 g is about 1/3 of the molar mass of KClO$_3$, and 0.5 × 2/3 = 1/3.**

# Practice

1. Phosphorus burns in air to produce a phosphorus oxide in the following reaction:

$$4P(s) + 5O_2(g) \rightarrow P_4O_{10}(s)$$

   **a.** What mass of phosphorus will be needed to produce 3.25 mol of P$_4$O$_{10}$?
   **ans: 403 g P**

   **b.** If 0.489 mol of phosphorus burns, what mass of oxygen is used? What mass of P$_4$O$_{10}$ is produced? **ans: 19.6 g O$_2$, 15.4 g P$_2$O$_4$**

2. Hydrogen peroxide breaks down, releasing oxygen, in the following reaction:

$$2H_2O_2(aq) \rightarrow 2H_2O(l) + O_2(g)$$

   **a.** What mass of oxygen is produced when 1.840 mol of H$_2$O$_2$ decomposes?
   **ans: 29.44 g O$_2$**

   **b.** What mass of water is produced when 5.0 mol O$_2$ is produced by this reaction? **ans: 180 g H$_2$O**

**Problem Solving** *continued*

## Sample Problem 3

**How many moles of aluminum will be produced from 30.0 kg $Al_2O_3$ in the following reaction?**

$$2Al_2O_3 \rightarrow 4Al + 3O_2$$

## Solution

### ANALYZE

What is given in the problem?     **the mass of aluminum oxide**

What are you asked to find?     **the amount of aluminum produced**

| Items | Data | |
|---|---|---|
| Substance | $Al_2O_3$ | Al |
| Coefficient in balanced equation | 2 | 4 |
| Molar mass | 101.96 g/mol | NA |
| Amount | ? mol | ? mol |
| Mass | 30.0 kg | NA |

### PLAN

What steps are needed to calculate the amount of Al produced from 30.0 kg of $Al_2O_3$?

**The molar mass of $Al_2O_3$ can be used to convert to moles $Al_2O_3$. The mole ratio of $Al:Al_2O_3$ from the coefficients in the equation will convert to moles Al from moles $Al_2O_3$.**

**1** Mass of $Al_2O_3$ in g ← *multiply by $\frac{1000\ g}{1\ kg}$* — Mass of $Al_2O_3$ in kg

*multiply by the inverse of the molar mass of $Al_2O_3$*

**2** Amount of $Al_2O_3$ in mol — *multiply by the mole ratio $\frac{4\ mol\ Al}{2\ mol\ Al_2O_3}$* → **3** Amount of Al in mol

$$\overset{given}{kg\ Al_2O_3} \times \frac{1000\ g}{kg} \times \frac{\overset{1}{\overline{molar\ mass\ Al_2O_3}}}{}\ \frac{1\ mol\ Al_2O_3}{101.96\ g\ Al_2O_3} \times \overset{mole\ ratio}{\frac{4\ mol\ Al}{2\ mol\ Al_2O_2}} = mol\ Al$$

### COMPUTE

$$30.0\ kg\ Al_2O_3 \times \frac{1000\ g}{kg} \times \frac{1\ mol\ Al_2O_3}{101.96\ g\ Al_2O_3} \times \frac{4\ mol\ Al}{2\ mol\ Al_2O_3} = 588\ mol\ Al$$

## EVALUATE

Are the units correct?
**Yes; units canceled to give moles of Al.**

Is the number of significant figures correct?
**Yes; three significant figures is correct.**

Is the answer reasonable?
**Yes; the molar mass of $Al_2O_3$ is about 100, so 30 kg of $Al_2O_3$ is about 300 mol. The mole ratio of $Al:Al_2O_3$ is 2:1, so the answer should be about 600 mol Al.**

# Practice

**1.** Sodium carbonate reacts with nitric acid according to the following equation.

$$Na_2CO_3(s) + 2HNO_3 \rightarrow 2NaNO_3 + CO_2 + H_2O$$

**a.** How many moles of $Na_2CO_3$ are required to produce 100.0 g of $NaNO_3$?
**ans: 0.5882 mol $Na_2CO_3$**

**b.** If 7.50 g of $Na_2CO_3$ reacts, how many moles of $CO_2$ are produced? **ans: 0.0708 mol $CO_2$**

**2.** Hydrogen is generated by passing hot steam over iron, which oxidizes to form $Fe_3O_4$, in the following equation.

$$3Fe(s) + 4H_2O(g) \rightarrow 4H_2(g) + Fe_3O_4(s)$$

**a.** If 625 g of $Fe_3O_4$ is produced in the reaction, how many moles of hydrogen are produced at the same time? **ans: 10.8 mol $H_2$**

**b.** How many moles of iron would be needed to generate 27 g of hydrogen? **ans: 10. mol Fe**

| Problem Solving *continued*

## Sample Problem 4

**Methane burns in air by the following reaction:**

$$CH_4(g) + 2O_2(g) \rightarrow CO_2(g) + 2H_2O(g)$$

**What mass of water is produced by burning 500. g of methane?**

## Solution

### ANALYZE

What is given in the problem?    **the mass of methane in grams**

What are you asked to find?    **the mass of water produced**

| Items | Data | |
|---|---|---|
| Substance | $CH_4$ | $H_2O$ |
| Coefficient in balanced equation | 1 | 2 |
| Molar mass | 16.05 g/mol | 18.02 g/mol |
| Amount | ? mol | ? mol |
| Mass | 500. g | ? g |

### PLAN

What steps are needed to calculate the mass of $H_2O$ produced from the burning of 500. g of $CH_4$?

**Convert grams of $CH_4$ to moles $CH_4$ by using the molar mass of $CH_4$. Use the mole ratio from the balanced equation to determine moles $H_2O$ from moles $CH_4$. Use the molar mass of $H_2O$ to calculate grams $H_2O$.**

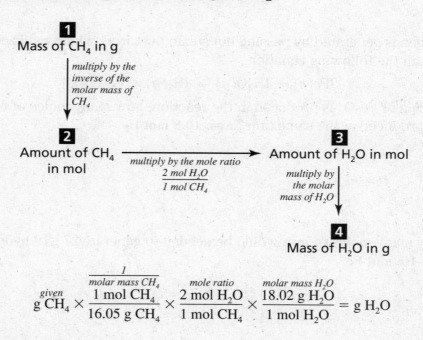

$$\underset{given}{g\ CH_4} \times \underset{\substack{molar\ mass\ CH_4}}{\frac{1\ mol\ CH_4}{16.05\ g\ CH_4}} \times \underset{mole\ ratio}{\frac{2\ mol\ H_2O}{1\ mol\ CH_4}} \times \underset{molar\ mass\ H_2O}{\frac{18.02\ g\ H_2O}{1\ mol\ H_2O}} = g\ H_2O$$

| Problem Solving *continued*

## COMPUTE

$$500. \text{ g } \cancel{CH_4} \times \frac{1 \text{ mol } \cancel{CH_4}}{16.05 \text{ g } \cancel{CH_4}} \times \frac{2 \text{ mol } \cancel{H_2O}}{1 \text{ mol } \cancel{CH_4}} \times \frac{18.02 \text{ g } H_2O}{1 \text{ mol } \cancel{H_2O}} = 1.12 \times 10^3 \text{ g } H_2O$$

## EVALUATE

Are the units correct?

**Yes; mass of $H_2O$ was required, and units canceled to give grams $H_2O$.**

Is the number of significant figures correct?

**Yes; three significant figures is correct because the mass of $CH_4$ was given to three significant figures.**

Is the answer reasonable?

**Yes; $CH_4$ and $H_2O$ have similar molar masses, and twice as many moles of $H_2O$ are produced as moles $CH_4$ burned. So, you would expect to get a little more than 1000 g of $H_2O$.**

# Practice

**1.** Calculate the mass of silver bromide produced from 22.5 g of silver nitrate in the following reaction:

$$2AgNO_3(aq) + MgBr_2(aq) \rightarrow 2AgBr(s) + Mg(NO_3)_2(aq) \text{ ans: } \textbf{24.9 g AgBr}$$

**2.** What mass of acetylene, $C_2H_2$, will be produced from the reaction of 90. g of calcium carbide, $CaC_2$, with water in the following reaction?

$$CaC_2(s) + 2H_2O(l) \rightarrow C_2H_2(g) + Ca(OH)_2(s) \text{ ans: } \textbf{37 g } C_2H_2$$

**3.** Chlorine gas can be produced in the laboratory by adding concentrated hydrochloric acid to manganese(IV) oxide in the following reaction:

$$MnO_2(s) + 4HCl(aq) \rightarrow MnCl_2(aq) + 2H_2O(l) + Cl_2(g)$$

**a.** Calculate the mass of $MnO_2$ needed to produce 25.0 g of $Cl_2$. **ans: 30.7 g $MnO_2$**

**b.** What mass of $MnCl_2$ is produced when 0.091 g of $Cl_2$ is generated? **ans: 0.16 g $MnCl_2$**

## Additional Problems

1. How many moles of ammonium sulfate can be made from the reaction of 30.0 mol of $NH_3$ with $H_2SO_4$ according to the following equation?

$$2NH_3 + H_2SO_4 \rightarrow (NH_4)_2SO_4$$

2. In a very violent reaction called a thermite reaction, aluminum metal reacts with iron(III) oxide to form iron metal and aluminum oxide according to the following equation:

$$Fe_2O_3 + 2Al \rightarrow 2Fe + Al_2O_3$$

   **a.** What mass of Al will react with 150 g of $Fe_2O_3$?

   **b.** If 0.905 mol $Al_2O_3$ is produced in the reaction, what mass of Fe is produced?

   **c.** How many moles of $Fe_2O_3$ will react with 99.0 g of Al?

3. As you saw in Sample Problem 1, the reaction $N_2(g) + 3H_2(g) \rightarrow 2NH_3(g)$ is used to produce ammonia commercially. If 1.40 g of $N_2$ are used in the reaction, how many grams of $H_2$ will be needed?

4. What mass of sulfuric acid, $H_2SO_4$, is required to react with 1.27 g of potassium hydroxide, KOH? The products of this reaction are potassium sulfate and water.

5. Ammonium hydrogen phosphate, $(NH_4)_2HPO_4$, a common fertilizer, is made from reacting phosphoric acid, $H_3PO_4$, with ammonia.

   **a.** Write the equation for this reaction.

   **b.** If 10.00 g of ammonia react, how many moles of fertilizer will be produced?

   **c.** What mass of ammonia will react with 2800 kg of $H_3PO_4$?

6. The following reaction shows the synthesis of zinc citrate, a ingredient in toothpaste, from zinc carbonate and citric acid.

$$3ZnCO_3(s) + 2C_6H_8O_7(aq) \rightarrow Zn_3(C_6H_5O_7)_2(aq) + 3H_2O(l) + 3CO_2(g)$$

   **a.** How many moles of $ZnCO_3$ and $C_6H_8O_7$ are required to produce 30.0 mol of $Zn_3(C_6H_5O_7)_2$?

   **b.** What quantities, in kilograms, of $H_2O$ and $CO_2$ are produced by the reaction of 500. mol of citric acid?

7. Methyl butanoate, an oily substance with a strong fruity fragrance, can be made by reacting butanoic acid with methanol according to the following equation:

$$C_3H_7COOH + CH_3OH \rightarrow C_3H_7COOCH_3 + H_2O$$

   **a.** What mass of methyl butanoate is produced from the reaction of 52.5 g of butanoic acid?

   **b.** In order to purify methyl butanoate, water must be removed. What mass of water is produced from the reaction of 5800. g of methanol?

**8.** Ammonium nitrate decomposes to yield nitrogen gas, water, and oxygen gas in the following reaction:

$$2NH_4NO_3 \rightarrow 2N_2 + O_2 + 4H_2O$$

**a.** How many moles of nitrogen gas are produced when 36.0 g of $NH_4NO_3$ reacts?

**b.** If 7.35 mol of $H_2O$ are produced in this reaction, what mass of $NH_4NO_3$ reacted?

**9.** Lead(II) nitrate reacts with potassium iodide to produce lead(II) iodide and potassium nitrate. If 1.23 mg of lead nitrate are consumed, what is the mass of the potassium nitrate produced?

**10.** A car battery produces electrical energy with the following chemical reaction:

$$Pb(s) + PbO_2(s) + 2H_2SO_4(aq) \rightarrow 2PbSO_4(s) + 2H_2O(l)$$

If the battery loses 0.34 kg of lead in this reaction, how many moles of lead(II) sulfate are produced?

**11.** In a space shuttle, the $CO_2$ that the crew exhales is removed from the air by a reaction within canisters of lithium hydroxide. On average, each astronaut exhales about 20.0 mol of $CO_2$ daily. What mass of water will be produced when this amount reacts with LiOH? The other product of the reaction is $Li_2CO_3$.

**12.** Water is sometimes removed from the products of a reaction by placing them in a closed container with excess $P_4O_{10}$. Water is absorbed by the following reaction:

$$P_4O_{10} + 6H_2O \rightarrow 4H_3PO_4$$

**a.** What mass of water can be absorbed by $1.00 \times 10^2$ g of $P_4O_{10}$?

**b.** If the $P_4O_{10}$ in the container absorbs 0.614 mol of water, what mass of $H_3PO_4$ is produced?

**c.** If the mass of the container of $P_4O_{10}$ increases from 56.64 g to 63.70 g, how many moles of water are absorbed?

**13.** Ethanol, $C_2H_5OH$, is considered a clean fuel because it burns in oxygen to produce carbon dioxide and water with few trace pollutants. If 95.0 g of $H_2O$ are produced during the combustion of ethanol, how many grams of ethanol were present at the beginning of the reaction?

**14.** Sulfur dioxide is one of the major contributors to acid rain. Sulfur dioxide can react with oxygen and water in the atmosphere to form sulfuric acid, as shown in the following equation:

$$2H_2O(l) + O_2(g) + 2SO_2(g) \rightarrow 2H_2SO_4(aq)$$

If 50.0 g of sulfur dioxide from pollutants reacts with water and oxygen found in the air, how many grams of sulfuric acid can be produced? How many grams of oxygen are used in the process?

**Problem Solving** *continued*

15. When heated, sodium bicarbonate, $NaHCO_3$, decomposes into sodium carbonate, $Na_2CO_3$, water, and carbon dioxide. If 5.00 g of $NaHCO_3$ decomposes, what is the mass of the carbon dioxide produced?

16. A reaction between hydrazine, $N_2H_4$, and dinitrogen tetroxide, $N_2O_4$, has been used to launch rockets into space. The reaction produces nitrogen gas and water vapor.

   **a.** Write a balanced chemical equation for this reaction.

   **b.** What is the mole ratio of $N_2O_4$ to $N_2$?

   **c.** How many moles of $N_2$ will be produced if 20 000 mol of $N_2H_4$ are used by a rocket?

   **d.** How many grams of $H_2O$ are made when 450. kg of $N_2O_4$ are consumed?

17. Joseph Priestley is credited with the discovery of oxygen. He produced $O_2$ by heating mercury(II) oxide, HgO, to decompose it into its elements. How many moles of oxygen could Priestley have produced if he had decomposed 517.84 g of mercury oxide?

18. Iron(III) chloride, $FeCl_3$, can be made by the reaction of iron with chlorine gas. How much iron, in grams, will be needed to completely react with 58.0 g of $Cl_2$?

19. Sodium sulfide and cadmium nitrate undergo a double-displacement reaction, as shown by the following equation:

$$Na_2S + Cd(NO_3)_2 \rightarrow 2NaNO_3 + CdS$$

What is the mass, in milligrams, of cadmium sulfide that can be made from 5.00 mg of sodium sulfide?

20. Potassium permanganate and glycerin react explosively according to the following equation:

$$14KMnO_4 + 4C_3H_5(OH)_3 \rightarrow 7K_2CO_3 + 7Mn_2O_3 + 5CO_2 + 16H_2O$$

   **a.** How many moles of carbon dioxide can be produced from 4.44 mol of $KMnO_4$?

   **b.** If 5.21 g of $H_2O$ are produced, how many moles of glycerin, $C_3H_5(OH)_3$, were used?

   **c.** If 3.39 mol of potassium carbonate are made, how many grams of manganese(III) oxide are also made?

   **d.** How many grams of glycerin will be needed to react with 50.0 g of $KMnO_4$? How many grams of $CO_2$ will be produced in the same reaction?

21. Calcium carbonate found in limestone and marble reacts with hydrochloric acid to form calcium chloride, carbon dioxide, and water according to the following equation:

$$CaCO_3(s) + 2HCl(aq) \rightarrow CaCl_2(aq) + CO_2(g) + H_2O(l)$$

   **a.** What mass of HCl will be needed to produce $5.00 \times 10^3$ kg of $CaCl_2$?

   **b.** What mass of $CO_2$ could be produced from the reaction of 750 g of $CaCO_3$?

**Problem Solving** *continued*

22. The fuel used to power the booster rockets on the space shuttle is a mixture of aluminum metal and ammonium perchlorate. The following balanced equation represents the reaction of these two ingredients:

$$3Al(s) + 3NH_4ClO_4(s) \rightarrow Al_2O_3(s) + AlCl_3(g) + 3NO(g) + 6H_2O(g)$$

**a.** If $1.50 \times 10^5$ g of Al react, what mass of $NH_4ClO_4$, in grams, is required?

**b.** If aluminum reacts with 620 kg of $NH_4ClO_4$, what mass of nitrogen monoxide is produced?

23. Phosphoric acid is typically produced by the action of sulfuric acid on rock that has a high content of calcium phosphate according to the following equation:

$$3H_2SO_4 + Ca_3(PO_4)_2 + 6H_2O \rightarrow 3[CaSO_4 \cdot 2H_2O] + 2H_3PO_4$$

**a.** If $2.50 \times 10^5$ kg of $H_2SO_4$ react, how many moles of $H_3PO_4$ can be made?

**b.** What mass of calcium sulfate dihydrate is produced by the reaction of 400. kg of calcium phosphate?

**c.** If the rock being used contains 78.8% $Ca_3(PO_4)_2$, how many metric tons of $H_3PO_4$ can be produced from 68 metric tons of rock?

24. Rusting of iron occurs in the presence of moisture according to the following equation:

$$4Fe(s) + 3O_2(g) \rightarrow 2Fe_2O_3(s)$$

Suppose that 3.19% of a heap of steel scrap with a mass of 1650 kg rusts in a year. What mass will the heap have after one year of rusting?

# Problem Solving

## Limiting Reactants

At the beginning of Chapter 8, a comparison was made between solving stoichiometry problems and making turkey sandwiches. Look at the sandwich recipe once more:

2 bread slices + 2 turkey slices + 1 lettuce leaf + 1 cheese slice →
$\qquad$ 1 turkey-and-cheese sandwich

If you have 24 slices of turkey, you can make 12 sandwiches at 2 slices per sandwich *if you have enough of all the other ingredients.* If, however, you have only 16 slices of bread, you can make only 8 sandwiches, even though you may an ample supply of the other ingredients. The bread is the *limiting* ingredient that prevents you from making more than 8 sandwiches.

The same idea applies to chemical reactions. Look at a reaction used to generate hydrogen gas in the laboratory:

$$Zn(s) + H_2SO_4(aq) \rightarrow ZnSO_4(aq) + H_2(g)$$

The balanced equation tells you that 1 mol Zn reacts with 1 mol $H_2SO_4$ to produce 1 mol $ZnSO_2$ and 1 mol $H_2$. Suppose you have 1 mol Zn and 5 mol $H_2SO_4$. What will happen, and what will you get? Only 1 mol of $H_2SO_4$ will react and only 1 mol of each of the products will be produced because only 1 mol Zn is available to react. In this situation, zinc is the limiting reactant. When it is used up the reaction stops even though more $H_2SO_4$ is available.

It is difficult to directly observe molar amounts of reactants as they are used up. It is much easier to determine when a certain mass of a reactant has been completely used. Use molar masses to restate the equation in terms of mass, as follows:

$$65.39 \text{ g Zn} + 98.09 \text{ g } H_2SO_4 \rightarrow 161.46 \text{ g } ZnSO_4 + 2.02 \text{ g } H_2$$

This version of the equation tells you that zinc and sulfuric acid will *always* react in a mass ratio of 65.39 g of Zn : 98.09 g of $H_2SO_4$ or 0.667 g of Zn : 1.000 g of $H_2SO_4$. If you have 65.39 g of Zn but only 87.55 g of $H_2SO_4$, you will not be able to make 2.02 g of hydrogen. Sulfuric acid will be the limiting reactant, preventing the zinc from reacting completely. Suppose you place 20 g of zinc and 100 g of sulfuric acid into a flask. Which would be used up first? In other words, is the limiting reactant zinc or sulfuric acid? How much of each product will be produced? The sample problems in this chapter will show you how to answer these questions.

**Problem Solving** *continued*

## General Plan for Solving Limiting Reactant Problems

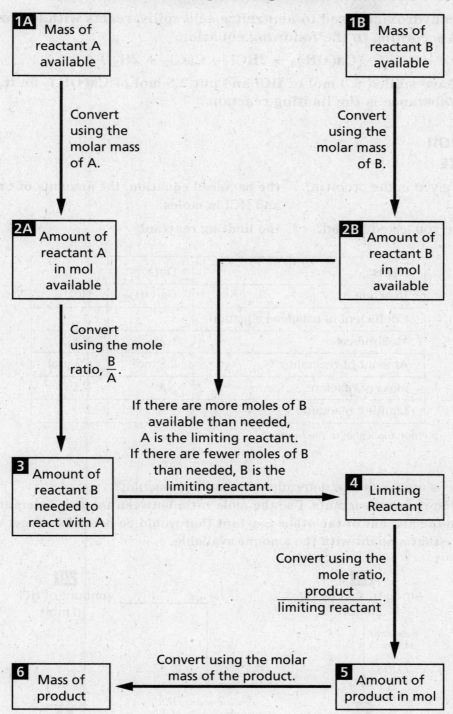

**1A** Mass of reactant A available

Convert using the molar mass of A.

**2A** Amount of reactant A in mol available

Convert using the mole ratio, $\frac{B}{A}$.

**3** Amount of reactant B needed to react with A

**1B** Mass of reactant B available

Convert using the molar mass of B.

**2B** Amount of reactant B in mol available

If there are more moles of B available than needed, A is the limiting reactant. If there are fewer moles of B than needed, B is the limiting reactant.

**4** Limiting Reactant

Convert using the mole ratio, $\frac{product}{limiting\ reactant}$.

**5** Amount of product in mol

Convert using the molar mass of the product.

**6** Mass of product

**| Problem Solving** *continued*

## Sample Problem 1

Calcium hydroxide, used to neutralize acid spills, reacts with hydrochloric acid according to the following equation:

$$Ca(OH)_2 + 2HCl \rightarrow CaCl_2 + 2H_2O$$

If you have spilled 6.3 mol of HCl and put 2.8 mol of $Ca(OH)_2$ on it, which substance is the limiting reactant?

## Solution

### ANALYZE

What is given in the problem?    **the balanced equation, the amounts of $Ca(OH)_2$ and HCl in moles**

What are you asked to find?    **the limiting reactant**

| Items | Data | |
|---|---|---|
| Reactant | $Ca(OH)_2$ | HCl |
| Coefficient in balanced equation | 1 | 2 |
| Molar mass | NA* | NA |
| Amount of reactant | 2.8 mol | 6.3 mol |
| Mass of reactant | NA | NA |
| Limiting reactant | ? | ? |

*not applicable to the problem

### PLAN

What steps are needed to determine the limiting reactant?
**Choose one of the reactants. Use the mole ratio between the two reactants to compute the amount of the other reactant that would be needed to react with it. Compare that amount with the amount available.**

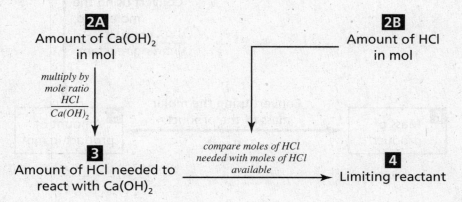

Choose one of the reactants, for instance, $Ca(OH)_2$

**Problem Solving** *continued*

$$\text{mol} \overset{given}{\text{Ca(OH)}_2} \times \frac{\overset{mole\ ratio}{2 \text{ mol HCl}}}{1 \text{ mol Ca(OH)}_2} = \text{mol HCl needed}$$

## COMPUTE

$$2.8 \text{ mol } \cancel{\text{Ca(OH)}_2} \times \frac{2 \text{ mol HCl}}{1 \text{ mol } \cancel{\text{Ca(OH)}_2}} = 5.6 \text{ mol HCl needed}$$

The computation shows that more HCl (6.3 mol) is available than is needed (5.6 mol) to react with the 2.8 mol $Ca(OH)_2$ available. Therefore, HCl is present in excess, making $Ca(OH)_2$ the limiting reactant.

## EVALUATE

Is the answer reasonable?

**Yes; you can see that 6.3 mol HCl is more than is needed to react with 2.8 mol $Ca(OH)_2$.**

# Practice

**1.** Aluminum oxidizes according to the following equation:

$$4Al + 3O_2 \rightarrow 2Al_2O_3$$

Powdered Al (0.048 mol) is placed into a container containing 0.030 mol $O_2$. What is the limiting reactant? **ans: $O_2$**

**| Problem Solving** *continued*

## Sample Problem 2

Chlorine can replace bromine in bromide compounds forming a chloride compound and elemental bromine. The following equation is an example of this reaction.

$$2KBr(aq) + Cl_2(aq) \rightarrow 2KCl(aq) + Br_2(l)$$

When 0.855 g of $Cl_2$ and 3.205 g of KBr are mixed in solution, which is the limiting reactant? How many grams of $Br_2$ are formed?

## Solution
### ANALYZE

What is given in the problem?        the balanced equation, and the masses of $Cl_2$ and KBr available

What are you asked to find?          which reactant is limiting, and the mass of $Br_2$ produced

| Items | Data | | |
|---|---|---|---|
| Substance | KBr | $Cl_2$ | $Br_2$ |
| Coefficient in balanced equation | 2 | 1 | 1 |
| Molar mass* | 119.00 g/mol | 70.90 g/mol | 159.80 g/mol |
| Amount of substance | ? mol | ? mol | ? mol |
| Mass of substance | 3.205 g | 0.855 g | ? g |
| Limiting reactant | ? | ? | NA |

*determined from the periodic table

### PLAN
What steps are needed to determine the limiting reactant?
**Convert mass of each reactant to amount in moles. Choose one of the reactants. Compute the amount of the other reactant needed. Compare that with the amount available.**

What steps are needed to determine the mass of $Br_2$ produced in the reaction?
**Use amount of the limiting reactant and the mole ratio given in the equation to determine the amount of $Br_2$. Convert the amount of $Br_2$ to the mass of $Br_2$ using the molar mass.**

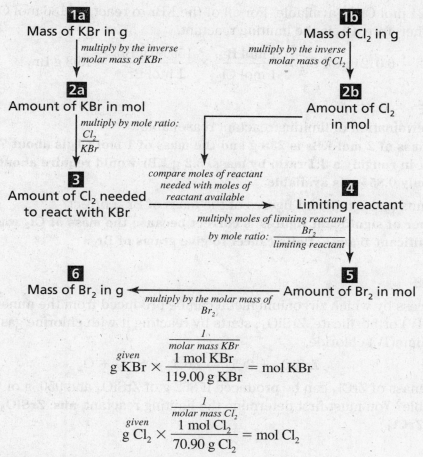

**1a** Mass of KBr in g
*multiply by the inverse molar mass of KBr*

**1b** Mass of $Cl_2$ in g
*multiply by the inverse molar mass of $Cl_2$*

**2a** Amount of KBr in mol
*multiply by mole ratio:* $\frac{Cl_2}{KBr}$

**2b** Amount of $Cl_2$ in mol

**3** Amount of $Cl_2$ needed to react with KBr

*compare moles of reactant needed with moles of reactant available*

**4** Limiting reactant
*multiply moles of limiting reactant by mole ratio:* $\frac{Br_2}{limiting\ reactant}$

**6** Mass of $Br_2$ in g ← *multiply by the molar mass of $Br_2$* — **5** Amount of $Br_2$ in mol

$$\overset{given}{g\ KBr} \times \frac{\overset{\frac{1}{molar\ mass\ KBr}}{1\ mol\ KBr}}{119.00\ g\ KBr} = mol\ KBr$$

$$\overset{given}{g\ Cl_2} \times \frac{\overset{\frac{1}{molar\ mass\ Cl_2}}{1\ mol\ Cl_2}}{70.90\ g\ Cl_2} = mol\ Cl_2$$

Choose one of the reactants, KBr for instance.

$$\overset{calculated\ above}{mol\ KBr} \times \frac{\overset{mole\ ratio}{1\ mol\ Cl_2}}{1\ mol\ KBr} = mol\ Cl_2\ needed$$

Determine the limiting reactant.

$$\overset{calculated\ above}{mol\ limiting\ reactant} \times \frac{\overset{mole\ ratio}{mol\ Br_2}}{mol\ limiting\ reactant} \times \frac{\overset{molar\ mass\ Br_2}{159.80\ g\ Br_2}}{1\ mol\ Br_2} = g\ Br_2$$

**COMPUTE**

$$3.205\ g\ KBr \times \frac{1\ mol\ KBr}{119.00\ g\ KBr} = 0.02693\ mol\ KBr$$

$$0.855\ g\ Cl_2 \times \frac{1\ mol\ Cl_2}{70.90\ g\ Cl_2} = 0.0121\ mol\ Cl_2$$

Choose one of the reactants, KBr, for instance.

$$0.02693\ mol\ KBr \times \frac{1\ mol\ Cl_2}{2\ mol\ KBr} = 0.01346\ mol\ Cl_2\ needed$$

Only 0.0121 mol $Cl_2$ is available. For all of the KBr to react, 0.0136 mol $Cl_2$ is needed. Therefore, $Cl_2$ is the limiting reactant.

$$0.0121 \text{ mol } Cl_2 \times \frac{1 \text{ mol } Br_2}{1 \text{ mol } Cl_2} \times \frac{159.80 \text{ g } Br_2}{1 \text{ mol } Br_2} = 1.93 \text{ g } Br_2$$

### EVALUATE

Is the determination of limiting reactant reasonable?

**Yes; the mass of 2 mol KBr is 238 g and the mass of 1 mol $Cl_2$ is about 71 g, so they react in roughly a 3:1 ratio by mass. 3.2 g KBr would require about 1 g of $Cl_2$, but only 0.855 g is available.**

Are the units and significant figures of the mass of $Br_2$ correct?

**The number of significant figures is correct because the mass of $Cl_2$ was given to three significant figures. Units cancel to give grams of $Br_2$.**

## Practice

1. A process by which zirconium metal can be produced from the mineral zirconium(IV) orthosilicate, $ZrSiO_4$, starts by reacting it with chlorine gas to form zirconium(IV) chloride.

$$ZrSiO_4 + 2Cl_2 \rightarrow ZrCl_4 + SiO_2 + O_2$$

What mass of $ZrCl_4$ can be produced if 862 g of $ZrSiO_4$ and 950. g of $Cl_2$ are available? You must first determine the limiting reactant. **ans: $ZrSiO_4$, 1.10 × $10^3$ g $ZrCl_4$**

# Additional Problems

**1.** Heating zinc sulfide in the presence of oxygen yields the following:

$$ZnS + O_2 \rightarrow ZnO + SO_2$$

If 1.72 mol of ZnS is heated in the presence of 3.04 mol of $O_2$, which reactant will be used up? Balance the equation first.

**2.** Use the following equation for the oxidation of aluminum in the following problems.

$$4Al + 3O_2 \rightarrow 2Al_2O_3$$

**a.** Which reactant is limiting if 0.32 mol Al and 0.26 mol $O_2$ are available?

**b.** How many moles of $Al_2O_3$ are formed from the reaction of $6.38 \times 10^{-3}$ mol of $O_2$ and $9.15 \times 10^{-3}$ mol of Al?

**c.** If 3.17 g of Al and 2.55 g of $O_2$ are available, which reactant is limiting?

**3.** In the production of copper from ore containing copper(II) sulfide, the ore is first roasted to change it to the oxide according to the following equation:

$$2CuS + 3O_2 \rightarrow 2CuO + 2SO_2$$

**a.** If 100 g of CuS and 56 g of $O_2$ are available, which reactant is limiting?

**b.** What mass of CuO can be formed from the reaction of 18.7 g of CuS and 12.0 g of $O_2$?

**4.** A reaction such as the one shown here is often used to demonstrate a single replacement reaction.

$$3CuSO_4(aq) + 2Fe(s) \rightarrow 3Cu(s) + Fe_2(SO_4)_3(aq)$$

If you place 0.092 mol of iron filings in a solution containing 0.158 mol of $CuSO_4$, what is the limiting reactant? How many moles of Cu will be formed?

**5.** In the reaction $BaCO_3 + 2HNO_3 \rightarrow Ba(NO_3)_2 + CO_2 + H_2O$, what mass of $Ba(NO_3)_2$ can be formed by combining 55 g $BaCO_3$ and 26 g $HNO_3$?

**6.** Bromine displaces iodine in magnesium iodide by the following process:

$$MgI_2 + Br_2 \rightarrow MgBr_2 + I_2$$

**a.** Which is the excess reactant when 560 g of $MgI_2$ and 360 g of $Br_2$ react, and what mass remains?

**b.** What mass of $I_2$ is formed in the same process?

**7.** Nickel displaces silver from silver nitrate in solution according to the following equation:

$$2AgNO_3 + Ni \rightarrow 2Ag + Ni(NO_3)_2$$

**a.** If you have 22.9 g of Ni and 112 g of $AgNO_3$, which reactant is in excess?

**b.** What mass of nickel(II) nitrate would be produced given the quantities above?

**8.** Carbon disulfide, $CS_2$, is an important industrial substance. Its fumes can burn explosively in air to form sulfur dioxide and carbon dioxide.

$$CS_2(g) + O_2(g) \rightarrow SO_2(g) + CO_2(g)$$

If 1.60 mol of $CS_2$ burns with 5.60 mol of $O_2$, how many moles of the excess reactant will still be present when the reaction is over?

**9.** Although poisonous, mercury compounds were once used to kill bacteria in wounds and on the skin. One was called "ammoniated mercury" and is made from mercury(II) chloride according to the following equation:

$$HgCl_2(aq) + 2NH_3(aq) \rightarrow Hg(NH_2)Cl(s) + NH_4Cl(aq)$$

**a.** What mass of $Hg(NH_2)Cl$ could be produced from 0.91 g of $HgCl_2$ assuming plenty of ammonia is available?

**b.** What mass of $Hg(NH_2)Cl$ could be produced from 0.91 g of $HgCl_2$ and 0.15 g of $NH_3$ in solution?

**10.** Aluminum chips are sometimes added to sodium hydroxide-based drain cleaners because they react to generate hydrogen gas which bubbles and helps loosen material in the drain. The equation follows.

$$Al(s) + NaOH(aq) + H_2O(l) \rightarrow NaAlO_2(aq) + H_2(g)$$

**a.** Balance the equation.

**b.** How many moles of $H_2$ can be generated from 0.57 mol Al and 0.37 mol NaOH in excess water?

**c.** Which reactant should be limiting in order for the mixture to be most effective as a drain cleaner? Explain your choice.

**11.** Copper is changed to copper(II) ions by nitric acid according to the following equation:

$$4HNO_3 + Cu \rightarrow Cu(NO_3)_2 + 2NO_2 + 2H_2O$$

**a.** How many moles each of $HNO_3$ and Cu must react in order to produce 0.0845 mol of $NO_2$?

**b.** If 5.94 g of Cu and 23.23 g of $HNO_3$ are combined, which reactant is in excess?

**12.** One industrial process for producing nitric acid begins with the following reaction:

$$4NH_3 + 5O_2 \rightarrow 4NO + 6H_2O$$

**a.** If 2.90 mol $NH_3$ and 3.75 mol $O_2$ are available, how many moles of each product are formed?

**b.** Which reactant is limiting if $4.20 \times 10^4$ g of $NH_3$ and $1.31 \times 10^5$ g of $O_2$ are available?

**c.** What mass of NO is formed in the reaction of 869 kg of $NH_3$ and 2480 kg $O_2$?

**13.** Acetaldehyde $CH_3CHO$ is manufactured by the reaction of ethanol with copper(II) oxide according to the following equation:

$$CH_3CH_2OH + CuO \rightarrow CH_3CHO + H_2O + Cu$$

What mass of acetaldehyde can be produced by the reaction between 620 g of ethanol and 1020 g of CuO? What mass of which reactant will be left over?

**14.** Hydrogen bromide can be produced by a reaction among bromine, sulfur dioxide, and water as follows.

$$SO_2 + Br_2 + H_2O \rightarrow 2HBr + H_2SO_4$$

If 250 g of $SO_2$ and 650 g of $Br_2$ react in the presence of excess water, what mass of HBr will be formed?

**15.** Sulfur dioxide can be produced in the laboratory by the reaction of hydrochloric acid and a sulfite salt such as sodium sulfite.

$$Na_2SO_3 + 2HCl \rightarrow 2NaCl + SO_2 + H_2O$$

What mass of $SO_2$ can be made from 25.0 g of $Na_2SO_3$ and 22.0 g of HCl?

**16.** The rare-earth metal terbium is produced from terbium(III) fluoride and calcium metal by the following displacement reaction:

$$2TbF_3 + 3Ca \rightarrow 3CaF_2 + 2Tb$$

**a.** Given 27.5 g of $TbF_3$ and 6.96 g of Ca, how many grams of terbium could be produced?

**b.** How many grams of the excess reactant are left over?

Skills Worksheet )

# Problem Solving

## Percentage Yield

Although we can write perfectly balanced equations to represent perfect reactions, the reactions themselves are often not perfect. A reaction does not always produce the quantity of products that the balanced equation seems to guarantee. This happens not because the equation is wrong but because reactions in the real world seldom produce perfect results.

As an example of an imperfect reaction, look again at the equation that shows the industrial production of ammonia.

$$N_2(g) + 3H_2(g) \rightarrow 2NH_3(g)$$

In the manufacture of ammonia, it is nearly impossible to produce 2 mol (34.08 g) of $NH_3$ from the simple reaction of 1 mol (28.02 g) of $N_2$ and 3 mol (6.06 g) of $H_2$ because some ammonia molecules begin breaking down into $N_2$ and $H_2$ molecules as soon as they are formed.

There are several reasons that real-world reactions do not produce products at a yield of 100%. Some are simple mechanical reasons, such as:

• Reactants or products leak out, especially when they are gases.
• The reactants are not 100% pure.
• Some product is lost when it is purified.

There are also many chemical reasons, including:

• The products decompose back into reactants (as with the ammonia process).
• The products react to form different substances.
• Some of the reactants react in ways other than the one shown in the equation. These are called *side reactions*.
• The reaction occurs very slowly. This is especially true of reactions involving organic substances.

Chemists are very concerned with the yields of reactions because they must find ways to carry out reactions economically and on a large scale. If the yield of a reaction is too small, the products may not be competitive in the marketplace. If a reaction has only a 50% yield, it produces only 50% of the amount of product that it theoretically should. In this chapter, you will learn how to solve problems involving real-world reactions and percentage yield.

## Problem Solving *continued*

### General Plan for Solving Percentage-Yield Problems

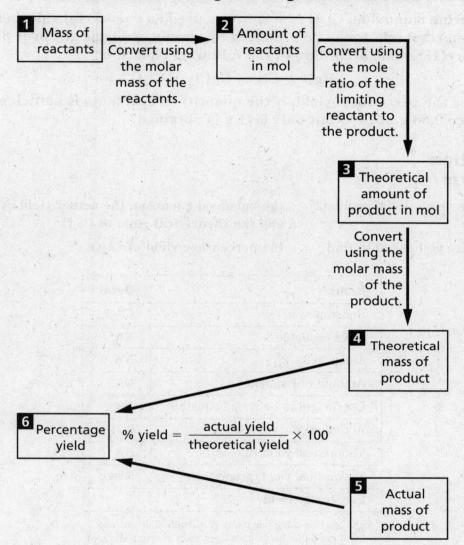

**1** Mass of reactants

Convert using the molar mass of the reactants.

**2** Amount of reactants in mol

Convert using the mole ratio of the limiting reactant to the product.

**3** Theoretical amount of product in mol

Convert using the molar mass of the product.

**4** Theoretical mass of product

**6** Percentage yield

$$\% \text{ yield} = \frac{\text{actual yield}}{\text{theoretical yield}} \times 100$$

**5** Actual mass of product

| **Problem Solving** *continued*

## Sample Problem 1

Dichlorine monoxide, $Cl_2O$ is sometimes used as a powerful chlorinating agent in research. It can be produced by passing chlorine gas over heated mercury(II) oxide according to the following equation:

$$HgO + Cl_2 \rightarrow HgCl_2 + Cl_2O$$

What is the percentage yield, if the quantity of reactants is sufficient to produce 0.86 g of $Cl_2O$ but only 0.71 g is obtained?

## Solution

### ANALYZE

What is given in the problem?     the balanced equation, the actual yield of $Cl_2O$, and the theoretical yield of $Cl_2O$

What are you asked to find?       the percentage yield of $Cl_2O$

| Items | Data |
|-------|------|
| Substance | $Cl_2O$ |
| Mass available | NA* |
| Molar mass | NA |
| Amount of reactant | NA |
| Coefficient in balanced equation | NA |
| Actual yield | 0.71 g |
| Theoretical yield (moles) | NA |
| Theoretical yield (grams) | 0.86 g |
| Percentage yield | ? |

\* Although this table has many *Not Applicable* entries, you will need much of this information in other kinds of percentage-yield problems.

### PLAN

What steps are needed to calculate the percentage yield of $Cl_2O$?

**Compute the ratio of the actual yield to the theoretical yield, and multiply by 100 to convert to a percentage.**

**Problem Solving** *continued*

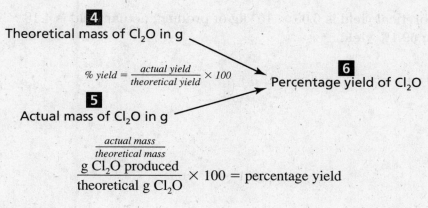

4  Theoretical mass of Cl₂O in g

$\% \text{ yield} = \dfrac{\text{actual yield}}{\text{theoretical yield}} \times 100$

6  Percentage yield of Cl₂O

5  Actual mass of Cl₂O in g

$$\dfrac{\frac{\text{actual mass}}{\text{theoretical mass}}}{\text{theoretical g Cl}_2\text{O}} \text{ g Cl}_2\text{O produced} \times 100 = \text{percentage yield}$$

## COMPUTE

$$\dfrac{0.71 \text{ g Cl}_2\text{O}}{0.86 \text{ g Cl}_2\text{O}} \times 100 = 83\% \text{ yield}$$

## EVALUATE

Are the units correct?
**Yes; the ratio was converted to a percentage.**

Is the number of significant figures correct?
**Yes; the number of significant figures is correct because the data were given to two significant figures.**

Is the answer reasonable?
**Yes; 83% is about 5/6, which appears to be close to the ratio 0.71/0.86.**

# Practice

1. Calculate the percentage yield in each of the following cases:

   **a.** theoretical yield is 50.0 g of product; actual yield is 41.9 g **ans: 83.8% yield**

   **b.** theoretical yield is 290 kg of product; actual yield is 270 kg **ans: 93% yield**

**Problem Solving** *continued*

**c.** theoretical yield is $6.05 \times 10^4$ kg of product; actual yield is $4.18 \times 10^4$ kg
**ans: 69.1% yield**

**d.** theoretical yield is 0.00192 g of product; actual yield is 0.00089 g **ans: 46% yield**

**❚ Problem Solving** *continued*

## Sample Problem 2

Acetylene, $C_2H_2$, can be used as an industrial starting material for the production of many organic compounds. Sometimes, it is first brominated to form 1,1,2,2-tetrabromoethane, $CHBr_2CHBr_2$, which can then be reacted in many different ways to make other substances. The equation for the bromination of acetylene follows:

$$\underset{acetylene}{C_2H_2} + 2Br_2 \rightarrow \underset{\text{1,1,2,2-tetrabromoethane}}{CHBr_2CHBR_2}$$

If 72.0 g of $C_2H_2$ reacts with excess bromine and 729 g of the product is recovered, what is the percentage yield of the reaction?

## Solution

### ANALYZE

What is given in the problem?     the balanced equation, the mass of acetylene
                                  that reacts, and the mass of tetrabromoethane
                                  produced

What are you asked to find?       the percentage yield of tetrabromoethane

| Items | Data | |
|---|---|---|
| Substance | $C_2H_2$ | $CHBr_2CHBr_2$ |
| Mass available | 72.0 g available | NA |
| Molar mass* | 26.04 g/mol | 345.64 g/mol |
| Amount of reactant | ? | NA |
| Coefficient in balanced equation | 1 | 1 |
| Actual yield | NA | 729 g |
| Theoretical yield (moles) | NA | ? |
| Theoretical yield (grams) | NA | ? |
| Percentage yield | NA | ? |

*determined from the periodic table

### PLAN

What steps are needed to calculate the theoretical yield of tetrabromoethane?
**Set up a stoichiometry calculation to find the amount of product that can be formed from the given amount of reactant.**

What steps are needed to calculate the percentage yield of tetrabromoethane?
**Compute the ratio of the actual yield to the theoretical yield, and multiply by 100 to convert to a percentage.**

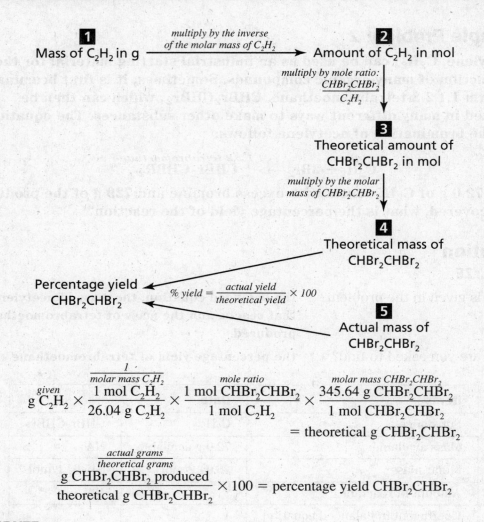

$$\underset{given}{\text{g C}_2\text{H}_2} \times \frac{\overset{\frac{1}{molar\ mass\ C_2H_2}}{1\ \text{mol C}_2\text{H}_2}}{26.04\ \text{g C}_2\text{H}_2} \times \frac{\overset{mole\ ratio}{1\ \text{mol CHBr}_2\text{CHBr}_2}}{1\ \text{mol C}_2\text{H}_2} \times \frac{\overset{molar\ mass\ CHBr_2CHBr_2}{345.64\ \text{g CHBr}_2\text{CHBr}_2}}{1\ \text{mol CHBr}_2\text{CHBr}_2}$$

$$= \text{theoretical g CHBr}_2\text{CHBr}_2$$

$$\frac{\overset{\frac{actual\ grams}{theoretical\ grams}}{\text{g CHBr}_2\text{CHBr}_2\ \text{produced}}}{\text{theoretical g CHBr}_2\text{CHBr}_2} \times 100 = \text{percentage yield CHBr}_2\text{CHBr}_2$$

**COMPUTE**

$$72.0\ \text{g C}_2\text{H}_2 \times \frac{1\ \text{mol C}_2\text{H}_2}{26.04\ \text{g C}_2\text{H}_2} \times \frac{1\ \text{mol CHBr}_2\text{CHBr}_2}{1\ \text{mol C}_2\text{H}_2} \times \frac{345.64\ \text{g CHBr}_2\text{CHBr}_2}{1\ \text{mol CHBr}_2\text{CHBr}_2}$$

$$= 956\ \text{g CHBr}_2\text{CHBr}_2$$

$$\frac{729\ \text{g CHBr}_2\text{CHBr}_2}{956\ \text{g CHBr}_2\text{CHBr}_2} \times 100 = 76.3\%\ \text{yield}$$

**EVALUATE**

Are the units correct?

**Yes; units canceled to give grams of CHBr₂CHBr₂. Also, the ratio was converted to a percentage.**

Is the number of significant figures correct?

**Yes; the number of significant figures is correct because the data were given to three significant figures.**

Is the answer reasonable?

**Yes; about 3 mol of acetylene were used and the theoretical yield is the mass of about 3 mol tetrabromoethane.**

**Problem Solving** *continued*

# Practice

1. In the commercial production of the element arsenic, arsenic(III) oxide is heated with carbon, which reduces the oxide to the metal according to the following equation:

$$2As_2O_3 + 3C \rightarrow 3CO_2 + 4As$$

**a.** If 8.87 g of $As_2O_3$ is used in the reaction and 5.33 g of As is produced, what is the percentage yield? **ans: 79.3% yield**

**b.** If 67 g of carbon is used up in a different reaction and 425 g of As is produced, calculate the percentage yield of this reaction. **ans: 76% yield**

# Additional Problems

1. Ethyl acetate is a sweet-smelling solvent used in varnishes and fingernail-polish remover. It is produced industrially by heating acetic acid and ethanol together in the presence of sulfuric acid, which is added to speed up the reaction. The ethyl acetate is distilled off as it is formed. The equation for the process is as follows.

$$\overset{acetic\ acid}{CH_3COOH} + \overset{ethanol}{CH_3CH_2OH} \overset{H_2SO_4}{\longrightarrow} \overset{ethyl\ acetate}{CH_3COOCH_2CH_3} + H_2O$$

Determine the percentage yield in the following cases:

**a.** 68.3 g of ethyl acetate should be produced but only 43.9 g is recovered.

**b.** 0.0419 mol of ethyl acetate is produced but 0.0722 mol is expected. (Hint: Percentage yield can also be calculated by dividing the actual yield in moles by the theoretical yield in moles.)

**c.** 4.29 mol of ethanol is reacted with excess acetic acid, but only 2.98 mol of ethyl acetate is produced.

**d.** A mixture of 0.58 mol ethanol and 0.82 mol acetic acid is reacted and 0.46 mol ethyl acetate is produced. (Hint: What is the limiting reactant?)

2. Assume the following hypothetical reaction takes place.

$$2A + 7B \rightarrow 4C + 3D$$

Calculate the percentage yield in each of the following cases:

**a.** The reaction of 0.0251 mol of A produces 0.0349 mol of C.

**b.** The reaction of 1.19 mol of A produces 1.41 mol of D.

**c.** The reaction of 189 mol of B produces 39 mol of D.

**d.** The reaction of 3500 mol of B produces 1700 mol of C.

3. Elemental phosphorus can be produced by heating calcium phosphate from rocks with silica sand ($SiO_2$) and carbon in the form of coke. The following reaction takes place.

$$Ca_3(PO_4)_2 + 3SiO_2 + 5C \rightarrow 3CaSiO_3 + 2P + 5CO$$

**a.** If 57 mol of $Ca_3(PO_4)_2$ is used and 101 mol of $CaSiO_3$ is obtained, what is the percentage yield?

**b.** Determine the percentage yield obtained if 1280 mol of carbon is consumed and 622 mol of $CaSiO_3$ is produced.

**c.** The engineer in charge of this process expects a yield of 81.5%. If $1.4 \times 10^5$ mol of $Ca_3(PO_4)_2$ is used, how many moles of phosphorus will be produced?

4. Tungsten (W) can be produced from its oxide by reacting the oxide with hydrogen at a high temperature according to the following equation:

$$WO_3 + 3H_2 \rightarrow W + 3H_2O$$

**a.** What is the percentage yield if 56.9 g of $WO_3$ yields 41.4 g of tungsten?

**b.** How many moles of tungsten will be produced from 3.72 g of $WO_3$ if the yield is 92.0%?

**c.** A chemist carries out this reaction and obtains 11.4 g of tungsten. If the percentage yield is 89.4%, what mass of $WO_3$ was used?

**5.** Carbon tetrachloride, $CCl_4$, is a solvent that was once used in large quantities in dry cleaning. Because it is a dense liquid that does not burn, it was also used in fire extinguishers. Unfortunately, its use was discontinued because it was found to be a carcinogen. It was manufactured by the following reaction:

$$CS_2 + 3Cl_2 \rightarrow CCl_4 + S_2Cl_2$$

The reaction was economical because the byproduct disulfur dichloride, $S_2Cl_2$, could be used by industry in the manufacture of rubber products and other materials.

**a.** What is the percentage yield of $CCl_4$ if 719 kg is produced from the reaction of 410. kg of $CS_2$.

**b.** If 67.5 g of $Cl_2$ are used in the reaction and 39.5 g of $S_2Cl_2$ is produced, what is the percentage yield?

**c.** If the percentage yield of the industrial process is 83.3%, how many kilograms of $CS_2$ should be reacted to obtain $5.00 \times 10^4$ kg of $CCl_4$? How many kilograms of $S_2Cl_2$ will be produced, assuming the same yield for that product?

**6.** Nitrogen dioxide, $NO_2$, can be converted to dinitrogen pentoxide, $N_2O_5$, by reacting it with ozone, $O_3$. The reaction of $NO_2$ takes place according to the following equation:

$$2NO_2(g) + O_3(g) \rightarrow N_2O_5(s \text{ or } g) + O_2(g)$$

**a.** Calculate the percentage yield for a reaction in which 0.38 g of $NO_2$ reacts and 0.36 g of $N_2O_5$ is recovered.

**b.** What mass of $N_2O_5$ will result from the reaction of 6.0 mol of $NO_2$ if there is a 61.1% yield in the reaction?

**7.** In the past, hydrogen chloride, HCl, was made using the *salt-cake* method as shown in the following equation:

$$2NaCl(s) + H_2SO_4(aq) \rightarrow Na_2SO_4(s) + 2HCl(g)$$

If 30.0 g of NaCl and 0.250 mol of $H_2SO_4$ are available, and 14.6 g of HCl is made, what is the percentage yield?

**8.** Cyanide compounds such as sodium cyanide, NaCN, are especially useful in gold refining because they will react with gold to form a stable compound that can then be separated and broken down to retrieve the gold. Ore containing only small quantities of gold can be used in this form of "chemical mining." The equation for the reaction follows.

$$4Au + 8NaCN + 2H_2O + O_2 \rightarrow 4NaAu(CN)_2 + 4NaOH$$

**a.** What percentage yield is obtained if 410 g of gold produces 540 g of $NaAu(CN)_2$?

**Problem Solving** *continued*

   **b.** Assuming a 79.6% yield in the conversion of gold to $NaAu(CN)_2$, what mass of gold would produce 1.00 kg of $NaAu(CN)_2$?

   **c.** Given the conditions in (b), what mass of gold ore that is 0.001% gold would be needed to produce 1.00 kg of $NaAu(CN)_2$?

**9.** Diiodine pentoxide is useful in devices such as respirators because it reacts with the dangerous gas carbon monoxide, CO, to produce relatively harmless $CO_2$ according to the following equation:

$$I_2O_5 + 5CO \rightarrow I_2 + 5CO_2$$

   **a.** In testing a respirator, 2.00 g of carbon monoxide gas is passed through diiodine pentoxide. Upon analyzing the results, it is found that 3.17 g of $I_2$ was produced. Calculate the percentage yield of the reaction.

   **b.** Assuming that the yield in (a) resulted because some of the CO did not react, calculate the mass of CO that passed through.

**10.** Sodium hypochlorite, NaClO, the main ingredient in household bleach, is produced by bubbling chlorine gas through a strong lye (sodium hydroxide, NaOH) solution. The following equation shows the reaction that occurs.

$$2NaOH(aq) + Cl_2(g) \rightarrow NaCl(aq) + NaClO(aq) + H_2O(l)$$

   **a.** What is the percentage yield of the reaction if 1.2 kg of $Cl_2$ reacts to form 0.90 kg of NaClO?

   **b.** If a plant operator wants to make 25 metric tons of NaClO per day at a yield of 91.8%, how many metric tons of chlorine gas must be on hand each day?

   **c.** What mass of NaCl is formed per mole of chlorine gas at a yield of 81.8%?

   **d.** At what rate in kg per hour must NaOH be replenished if the reaction produces 370 kg/h of NaClO at a yield of 79.5%? Assume that all of the NaOH reacts to produce this yield.

**11.** Magnesium burns in oxygen to form magnesium oxide. However, when magnesium burns in air, which is only about 1/5 oxygen, side reactions form other products, such as magnesium nitride, $Mg_3N_2$.

   **a.** Write a balanced equation for the burning of magnesium in oxygen.

   **b.** If enough magnesium burns in air to produce 2.04 g of magnesium oxide but only 1.79 g is obtained, what is the percentage yield?

   **c.** Magnesium will react with pure nitrogen to form the nitride, $Mg_3N_2$. Write a balanced equation for this reaction.

   **d.** If 0.097 mol of Mg react with nitrogen and 0.027 mol of $Mg_3N_2$ is produced, what is the percentage yield of the reaction?

**| Problem Solving** *continued*

12. Some alcohols can be converted to organic acids by using sodium dichromate and sulfuric acid. The following equation shows the reaction of 1-propanol to propanoic acid.

$$3CH_3CH_2CH_2OH + 2Na_2Cr_2O_7 + 8H_2SO_4 \rightarrow$$
$$3CH_3CH_2COOH + 2Cr_2(SO_4)_3 + 2Na_2SO_4 + 11H_2O$$

   **a.** If 0.89 g of 1-propanol reacts and 0.88 g of propanoic acid is produced, what is the percentage yield?

   **b.** A chemist uses this reaction to obtain 1.50 mol of propanoic acid. The reaction consumes 136 g of propanol. Calculate the percentage yield.

   **c.** Some 1-propanol of uncertain purity is used in the reaction. If 116 g of $Na_2Cr_2O_7$ are consumed in the reaction and 28.1 g of propanoic acid are produced, what is the percentage yield?

13. Acrylonitrile, $C_3H_3N(g)$, is an important ingredient in the production of various fibers and plastics. Acrylonitrile is produced from the following reaction:

$$C_3H_6(g) + NH_3(g) + O_2(g) \rightarrow C_3H_3N(g) + H_2O(g)$$

   If 850. g of $C_3H_6$ is mixed with 300. g of $NH_3$ and unlimited $O_2$, to produce 850. g of acrylonitrile, what is the percentage yield? You must first balance the equation.

14. Methanol, $CH_3OH$, is frequently used in race cars as fuel. It is produced as the sole product of the combination of carbon monoxide gas and hydrogen gas.

   **a.** If 430. kg of hydrogen react, what mass of methanol could be produced?

   **b.** If $3.12 \times 10^3$ kg of methanol are actually produced, what is the percentage yield?

15. The compound, $C_6H_{16}N_2$, is one of the starting materials in the production of nylon. It can be prepared from the following reaction involving adipic acid, $C_6H_{10}O_4$:

$$C_6H_{10}O_4(l) + 2NH_3(g) + 4H_2(g) \rightarrow C_6H_{16}N_2(l) + 4H_2O(l)$$

   What is the percentage yield if 750. g of adipic acid results in the production of 578 g of $C_6H_{16}N_2$?

16. Plants convert carbon dioxide to oxygen during photosynthesis according to the following equation:

$$CO_2 + H_2O \rightarrow C_6H_{12}O_6 + O_2$$

   Balance this equation, and calculate how much oxygen would be produced if $1.37 \times 10^4$ g of carbon dioxide reacts with a percentage yield of 63.4%.

17. Lime, CaO, is frequently added to streams and lakes which have been polluted by acid rain. The calcium oxide reacts with the water to form a base that can neutralize the acid as shown in the following reaction:

$$CaO(s) + H_2O(l) \rightarrow Ca(OH)_2(s)$$

   If $2.67 \times 10^2$ mol of base are needed to neutralize the acid in a lake, and the above reaction has a percentage yield of 54.3%, what is the mass, in kilograms, of lime that must be added to the lake?

# Problem Solving

## Thermochemistry

Thermochemistry deals with the changes in energy that accompany a chemical reaction. Energy is measured in a quantity called *enthalpy*, represented as $H$. The change in energy that accompanies a chemical reaction is represented as $\Delta H$. Hess's law provides a method for calculating the $\Delta H$ of a reaction from tabulated data. This law states that if two or more chemical equations are added, the $\Delta H$ of the individual equations may also be added to find the $\Delta H$ of the final equation. As an example of how this law operates, look at the three reactions below.

| | | |
|---|---|---|
| (1) | $2H_2(g) + O_2(g) \rightarrow 2H_2O(l)$ | $\Delta H = -571.6$ kJ/mol |
| (2) | $2H_2O_2(l) \rightarrow 2H_2(g) + 2O_2(g)$ | $\Delta H = +375.6$ kJ/mol |
| (3) | $2H_2O_2(l) \rightarrow 2H_2O(l) + O_2(g)$ | $\Delta H = ?$ kJ/mol |

When adding equations 1 and 2, the 2 mol of $H_2(g)$ will cancel each other out, while only 1 mol of $O_2(g)$ will cancel.

$$2\cancel{H_2(g)} + \cancel{O_2(g)} \rightarrow 2H_2O(l)$$

$$2H_2O_2(l) \rightarrow 2\cancel{H_2(g)} + \overset{1}{\cancel{2}}O_2(g)$$

Combining what is left yields the following equation.

$$2H_2O_2(l) \rightarrow 2H_2O(l) + O_2(g)$$

Notice that this is the same equation as the third equation shown above. Adding the two $\Delta H$ values for the reactions 1 and 2 gives the $\Delta H$ value for reaction 3. Using Hess's law to calculate the enthalpy of this reaction, the following answer is obtained.

$$-571.6 \text{ kJ/mol} + 375.6 \text{ kJ/mol} = -196.0 \text{ kJ/mol}$$

Thus, the $\Delta H$ value for the reaction is $-196.0$ kJ/mol.

Equation 1 represents the formation of water from its elemental components. If equation 2 were written in reverse, it would represent the formation of hydrogen peroxide from its elemental components. Therefore, adding equations 1 and 2 is the equivalent of subtracting the equation for the formation of the reactants of equation 3 from the equation for the formation of the products of equation 3.

$$2H_2(g) + O_2(g) \rightarrow 2H_2O(l)$$
$$\underline{-[2H_2(g) + 2O_2(g) \rightarrow 2H_2O_2(l)]}$$
$$2H_2O_2(l) \rightarrow 2H_2O(l) + O_2(g)$$

The enthalpy of the final reaction can be rewritten using the following equation.

$$\Delta H_{\text{reaction}} = \text{sum of } \Delta H^0_{f_{\text{products}}} - \text{sum of } \Delta H^0_{f_{\text{reactants}}}$$

The equation states that the enthalpy change of a reaction is equal to the sum of the enthalpies of formation of the products minus the sum of the enthalpies of formation of the reactants. This allows Hess's law to be extended to state that the

**Problem Solving** *continued*

enthalpy change of any reaction can be calculated by looking up the standard molar enthalpy of formation, $\Delta H_f^0$, of each substance involved. Some common enthalpies of formation may be found in **Table 1.**

The enthalpy change, however, does not account for all of the energy change of a reaction. Changes in the disorder (entropy) of a system can add to or detract from the energy involved in the enthalpy change. This amount of energy is given by the expression $T\Delta S$, where $T$ is the Kelvin temperature and $\Delta S$ is the change in entropy during the reaction. A large increase in entropy, such as when a gas is produced from a reaction of liquids or solids, can contribute significantly to the overall energy change. The total amount of energy available from a reaction is called *Gibbs energy* and is denoted by $\Delta G$. Gibbs energy is given by the following equation.

$$\Delta G_{\text{reaction}} = \Delta H_{\text{reaction}} - T\Delta S_{\text{reaction}}$$

## TABLE 1 STANDARD ENTHALPIES OF FORMATION

| Substance | $\Delta H_f^0$ (kJ/mol) | Substance | $\Delta H_f^0$ (kJ/mol) |
|---|---|---|---|
| $NH_3(g)$ | −45.9 | $HF(g)$ | −273.3 |
| $NH_4Cl(s)$ | −314.4 | $H_2O(g)$ | −241.82 |
| $NH_4F(s)$ | −125 | $H_2O(l)$ | −285.8 |
| $NH_4NO_3(s)$ | −365.56 | $H_2O_2(l)$ | −187.8 |
| $Br_2(l)$ | 0.00 | $H_2SO_4(l)$ | −813.989 |
| $CaCO_3(s)$ | −1207.6 | $FeO(s)$ | −825.5 |
| $CaO(s)$ | −634.9 | $Fe_2O_3(s)$ | −1118.4 |
| $CH_4(g)$ | −74.9 | $MnO_2(s)$ | −520.0 |
| $C_3H_8(g)$ | −104.7 | $N_2O(g)$ | +82.1 |
| $CO_2(g)$ | −393.5 | $O_2(g)$ | 0.00 |
| $F_2(g)$ | 0.00 | $Na_2O(s)$ | −414.2 |
| $H_2(g)$ | 0.00 | $Na_2SO_3(s)$ | −1101 |
| $HBr(g)$ | −36.29 | $SO_2(g)$ | −296.8 |
| $HCl(g)$ | −92.3 | $SO_3(g)$ | −395.7 |

## Problem Solving *continued*

### General Plan for Solving Thermochemistry Problems

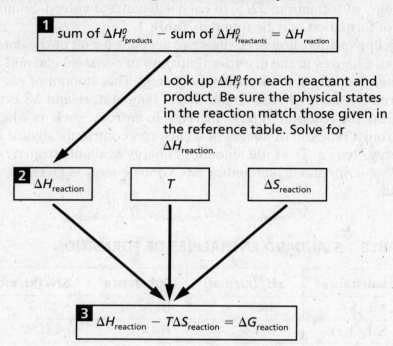

**1** sum of $\Delta H^0_{f\text{products}}$ − sum of $\Delta H^0_{f\text{reactants}}$ = $\Delta H_{\text{reaction}}$

Look up $\Delta H^0_f$ for each reactant and product. Be sure the physical states in the reaction match those given in the reference table. Solve for $\Delta H_{\text{reaction}}$.

**2** $\Delta H_{\text{reaction}}$     $T$     $\Delta S_{\text{reaction}}$

**3** $\Delta H_{\text{reaction}} - T\Delta S_{\text{reaction}} = \Delta G_{\text{reaction}}$

**Problem Solving** *continued*

## Sample Problem 1

Given the following two reactions and enthalpy data, calculate the enthalpy change for the reaction in which methane and oxygen combine to form ketene, $CH_2CO$, and water.

$$CH_2CO(g) + 2O_2(g) \rightarrow 2CO_2(g) + H_2O(g) \qquad \Delta H = -981.1 \text{ kJ}$$
$$CH_4(g) + 2O_2(g) \rightarrow CO_2(g) + 2H_2O(g) \qquad \Delta H = -802.3 \text{ kJ}$$

## Solution

### ANALYZE

What is given in the problem?    the desired product, chemical equations that can be added to obtain the desired product, and enthalpy changes for these chemical equations

What are you asked to find?    enthalpy change for the reaction in which methane and oxygen combine to form ketene and water

| Items | Data |
|-------|------|
| $\Delta H$ for reaction 1 | $-981.1$ kJ |
| $\Delta H$ for reaction 2 | $-802.3$ kJ |

### PLAN

What steps are needed to calculate $\Delta H$ for the reaction between methane and oxygen to form $CH_2CO$?

First, the two equations must be added to produce the final reaction. The first equation must be reversed so that ketene is a product, as shown in the final equation. The second equation must be multiplied by 2 so that carbon dioxide cancels out of the final equation. Then the individual enthalpies for the reactions must be added, adjusting for the fact that equation 1 is reversed and equation 2 is doubled.

$$2CO_2(g) + H_2O(g) \rightarrow CH_2CO(g) + 2O_2(g)$$
$$\underline{2 \times [CH_4(g) + 2O_2(g) \rightarrow CO_2(g) + 2H_2O(g)]}$$
$$2CH_4(g) + 2O_2(g) \rightarrow CH_2CO(g) + 3H_2O(g)$$

$$-\Delta H_{\text{reaction 1}}$$
$$\underline{+(2 \times \Delta H_{\text{reaction 2}})}$$
$$\Delta H_{\text{final reaction}}$$

**| Problem Solving** *continued*

## COMPUTE

$$2CO_2(g) + H_2O(g) \rightarrow CH_2CO(g) + 2O_2(g)$$
$$2CH_4(g) + 4O_2(g) \rightarrow 2CO_2(g) + 4H_2O(g)$$

$$\overline{2CO_2(g)} + \overline{H_2O(g)} + 2CH_4(g) + \overset{2}{\cancel{4}}O_2(g) \rightarrow$$
$$CH_2CO(g) + 2O_2(g) + 2CO_2(g) + \overset{3}{\cancel{4}}H_2O(g)$$

$$2CH_4(g) + 2O_2(g) \rightarrow CH_2CO(g) + 3H_2O(g)$$

$$\begin{array}{r} -(-981.1 \text{ kJ}) \\ +(2 \times -802.3 \text{ kJ}) \\ \hline -623.5 \text{ kJ} \end{array}$$

## EVALUATE

Are the units correct?
**Yes; adding terms in kilojoules gives an answer in kilojoules.**

Is the number of significant figures correct?
**Yes; the significant figures are correct. Rules for adding and rounding measurements give a result to four significant figures.**

Is the answer reasonable?
**Yes; the result can be approximated as −1600 kJ + 1000 kJ = −600 kJ.**

# Practice

**1.** Calculate the reaction enthalpy for the following reaction.

$$5CO_2(g) + Si_3N_4(s) \rightarrow 3SiO(s) + 2N_2O(g) + 5CO(g)$$

Use the following equations and data.

$$(1) \quad CO(g) + SiO_2(s) \rightarrow SiO(g) + CO_2(g)$$
$$(2) \quad 8CO_2(g) + Si_3N_4(s) \rightarrow 3SiO_2(s) + 2N_2O(g) + 8CO(g)$$

$\Delta H_{\text{reaction 1}} = +520.9 \text{ kJ}$
$\Delta H_{\text{reaction 2}} = +461.05 \text{ kJ}$ **ans: 2024 kJ**

## Sample Problem 2

**Calculate the enthalpy of reaction for the decomposition of hydrogen peroxide to water and oxygen gas according to the following equation.**

$$H_2O_2(l) \rightarrow H_2O(l) + O_2(g)$$

**Use Table 1 for the necessary enthalpies of formation.**

## Solution

### ANALYZE

What is given in the problem?  **the equation for the decomposition of $H_2O_2$, enthalpies of formation given in Table 1**

What are you asked to find?  **enthalpy of reaction for the decomposition of $H_2O_2$**

| Items | Data |
|---|---|
| $\Delta H$ decomposition of $H_2O_2(l)$ | ? kJ/mol |
| $\Delta H_f^0\, H_2O_2(l)$ | −187.8 kJ/mol* |
| $\Delta H_f^0\, O_2(g)$ | 0.00 kJ/mol** |
| $\Delta H_f^0\, H_2O(l)$ | −285.8 kJ/mol* |

\* from Table 1
\*\* any pure elemental substance has a $\Delta H_f^0$ of zero

### PLAN

What steps are needed to calculate $\Delta H$ for the decomposition reaction?
**First, the equation must be balanced. Add up the enthalpies of formation for the products. From this quantity, subtract the enthalpy of formation for the reactant.**

Balance the equation for the decomposition of hydrogen peroxide.

$$2H_2O_2(l) \rightarrow 2H_2O(l) + O_2(g)$$

**1**

$$\text{sum of } \Delta H_{f\,products}^0 - \text{sum of } \Delta H_{f\,reactants}^0 = \Delta H_{reaction}$$

*look up $\Delta H_f^0$ for each reactant and product, and solve*

**2**

$$\Delta H_{reaction}$$

*given in Table 1*

$$\left(2\Delta H_{f_{H_2O}}^0 + \Delta H_{f_{O_2}}^0\right) - \left(2\Delta H_{f_{H_2O_2}}^0\right) = \Delta H_{reaction}$$

### COMPUTE

$$[2(-285.8 \text{ kJ/mol}) + 0.00 \text{ kJ/mol}] - [2(-187.8 \text{ kJ/mol})] = -196.0 \text{ kJ/mol}$$

**| Problem Solving** *continued*

**EVALUATE**

Are the units correct?

**Yes; adding terms in kJ/mol gives kJ/mol.**

Is the number of significant figures correct?

**Yes; rules for adding and rounding measurements give a result to four significant figures.**

Is the answer reasonable?

**Yes; the result can be approximated as −600 kJ/mol + 400 kJ/mol = −200 kJ/mol.**

## Practice

**Determine ΔH for each of the following reactions.**

1. The following reaction is used to make CaO from limestone.

$$CaCO_3(s) \rightarrow CaO(s) + CO_2(g) \text{ ans: 179.2 kJ/mol}$$

2. The following reaction represents the oxidation of FeO to $Fe_2O_3$.

$$2FeO(s) + O_2(g) \rightarrow Fe_2O_3(s) \text{ ans: 533 kJ/mol}$$

3. The following reaction of ammonia and hydrogen fluoride produces ammonium fluoride.

$$NH_3(g) + HF(g) \rightarrow NH_4F(s) \text{ ans: 194 kJ/mol}$$

**| Problem Solving** *continued*

# Sample Problem 3

**Calculate the Gibbs energy change for the following reaction at 25°C.**

$$Ca(s) + 2H_2O(l) \rightarrow Ca(OH)_2(s) + H_2(g)$$

**Use the following data.**

$$\Delta H_{reaction} = -411.6 \text{ kJ/mol}; \Delta S_{reaction} = 31.8 \text{ J/mol·K}$$

# Solution

## ANALYZE

What is given in the problem?    **T, ΔH, and ΔS for the reaction**

What are you asked to find?    **the Gibbs energy of the reaction, ΔG**

| Items | Data |
|---|---|
| $\Delta H$ for the reaction at 25°C | −411.6 kJ/mol |
| $\Delta S$ for the reaction at 25°C | 31.8 J/mol·K |
| Temperature | 25°C = 298 K |
| $\Delta G$ for the reaction at 25°C | ? kJ/mol |

## PLAN

What steps are needed to calculate $\Delta G$ for the given reaction?

**Apply the relationship $\Delta G = \Delta H - T\Delta S$.**

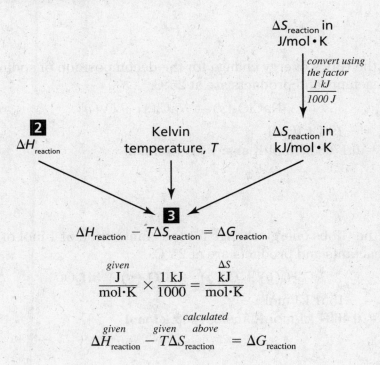

$$\Delta H_{reaction} - T\Delta S_{reaction} = \Delta G_{reaction}$$

$$\overset{given}{\frac{J}{mol \cdot K}} \times \frac{1 \text{ kJ}}{1000} = \overset{\Delta S}{\frac{kJ}{mol \cdot K}}$$

$$\overset{given}{\Delta H_{reaction}} - \overset{given}{T}\overset{calculated}{\Delta S_{reaction}} = \Delta G_{reaction}$$

### COMPUTE

$$\frac{31.8 \text{ J}}{\text{mol·K}} \times \frac{1 \text{ kJ}}{1000 \text{ J}} = 0.0318 \text{ kJ/mol·K}$$

$$-411.1 \text{ kJ/mol} - (298 \text{ K} \times 0.0318 \text{ kJ/mol·K}) = -411.6 \text{ kJ/mol} - 9.5 \text{ kJ/mol}$$

$$\Delta G_{\text{reaction}} = -421.1 \text{ kJ/mol}$$

Are the units correct?

**Yes; the kelvin units canceled. Adding terms in kJ/mol gives kJ/mol.**

Is the number of significant figures correct?

**Yes; the number of significant figures is correct. Rules for adding and rounding measurements give a result with four significant figures.**

Is the answer reasonable?

**Yes; values were given for $T$, $\Delta H$, and $\Delta S$ and the computation was carried out correctly.**

## Practice

**1.** Calculate the Gibbs energy change, $\Delta G$, for the combustion of hydrogen sulfide according to the following chemical equation. Assume reactants and products are at 25°C.

$$H_2S(g) + O_2(g) \rightarrow H_2O(l) + SO_2(g)$$

$\Delta H_{\text{reaction}} = -562.1 \text{ kJ/mol}$
$\Delta S_{\text{reaction}} = -0.09278 \text{ kJ/mol·K}$ **ans: −534.5 kJ/mol**

**2.** Calculate the Gibbs energy change for the decomposition of sodium chlorate. Assume reactants and products are at 25°C.

$$NaClO_3(s) \rightarrow NaCl(s) + O_2(g)$$

$\Delta H_{\text{reaction}} = -19.1 \text{ kJ/mol}$
$\Delta S_{\text{reaction}} = 0.1768 \text{ kJ/mol·K}$ **ans: −71.8 kJ/mol**

**3.** Calculate the Gibbs energy change for the combustion of 1 mol of ethane. Assume reactants and products are at 25°C.

$$C_2H_6(g) + O_2(g) \rightarrow 2CO_2(g) + 3H_2O(l)$$

$\Delta H_{\text{reaction}} = -1561 \text{ kJ/mol}$
$\Delta S_{\text{reaction}} = 0.4084 \text{ kJ/mol·K}$ **ans: −1683 kJ/mol**

**Problem Solving** *continued*

# Additional Problems

1. Calculate $\Delta H$ for the violent reaction of fluorine with water.

$$F_2(g) + H_2O(l) \rightarrow 2HF(g) + O_2(g)$$

2. Calculate $\Delta H$ for the reaction of calcium oxide and sulfur trioxide.

$$CaO(s) + SO_3(g) \rightarrow CaSO_4(s)$$

Use the following equations and data.

| | |
|---|---|
| $H_2O(l) + SO_3(g) \rightarrow H_2SO_4(l)$ | $\Delta H = -132.5 \text{ kJ/mol}$ |
| $H_2SO_4(l) + Ca(s) \rightarrow CaSO_4(s) + H_2(g)$ | $\Delta H = -602.5 \text{ kJ/mol}$ |
| $Ca(s) + O_2(g) \rightarrow CaO(s)$ | $\Delta H = -634.9 \text{ kJ/mol}$ |
| $H_2(g) + O_2(g) \rightarrow H_2O(l)$ | $\Delta H = -285.8 \text{ kJ/mol}$ |

3. Calculate $\Delta H$ for the reaction of sodium oxide with sulfur dioxide.

$$Na_2O(s) + SO_2(g) \rightarrow Na_2SO_3(s)$$

4. Use enthalpies of combustion to calculate $\Delta H$ for the oxidation of 1-butanol to make butanoic acid.

$$C_4H_9OH(l) + O_2(g) \rightarrow C_3H_7COOH(l) + H_2O(l)$$

Combustion of butanol:

$$C_4H_9OH(l) + 6O_2(g) \rightarrow 4CO_2(g) + 5H_2O(l)$$

$\Delta H_c = -2675.9 \text{ kJ/mol}$

Combustion of butanoic acid:

$$C_3H_7COOH(l) + 5O_2(g) \rightarrow 4CO_2(g) + 4H_2O(l)$$

$\Delta H_c = -2183.6 \text{ kJ/mol}$

5. Determine the free energy change for the reduction of CuO with hydrogen. Products and reactants are at 25°C.

$$CuO(s) + H_2(g) \rightarrow Cu(s) + H_2O(l)$$

$\Delta H = -128.5 \text{ kJ/mol}$
$\Delta S = -70.1 \text{ J/mol·K}$

6. Calculate the enthalpy change at 25°C for the reaction of sodium iodide and chlorine. Use only the data given.

$$NaI(s) + Cl_2(g) \rightarrow NaCl(s) + I_2(l)$$

$\Delta S = -79.9 \text{ J/mol·K}$
$\Delta G = -98.0 \text{ kJ/mol}$

7. The element bromine can be produced by the reaction of hydrogen bromide and manganese(IV) oxide.

$$4HBr(g) + MnO_2(s) \rightarrow MnBr_2(s) + 2H_2O(l) + Br_2(l)$$

$\Delta H$ for the reaction is $-291.3$ kJ/mol at 25°C. Use this value and values of $\Delta H_f^0$ from **Table 1** to calculate $\Delta H_f^0$ of $MnBr_2(s)$.

**| Problem Solving** *continued*

---

**8.** Calculate the change in entropy, $\Delta S$, at 25°C for the reaction of calcium carbide with water to produce acetylene gas.

$$CaC_2(s) + 2H_2O(l) \rightarrow C_2H_2(g) + Ca(OH)_2(s)$$

$\Delta G = -147.7$ kJ/mol
$\Delta H = -125.6$ kJ/mol

**9.** Calculate the Gibbs energy change for the explosive decomposition of ammonium nitrate at 25°C. Note that $H_2O$ is a gas in this reaction.

$$NH_4NO_3(s) \rightarrow N_2O(g) + 2H_2O(g)$$

$\Delta S = 446.4$ J/mol·K

**10.** In locations where natural gas, which is mostly methane, is not available, many people burn propane, which is delivered by truck and stored in a tank under pressure.

**a.** Write the chemical equations for the complete combustion of 1 mol of methane, $CH_4$, and 1 mol of propane, $C_3H_8$.

**b.** Calculate the enthalpy change for each reaction to determine the amount of energy released off by burning 1 mol of each fuel.

**c.** Using the enthapies of combustion you calculated, determine the energy output per kilogram of each fuel. Which fuel yields more energy per unit mass?

**11.** The hydration of acetylene to form acetaldehyde is shown in the following equation:

$$C_2H_2(g) + H_2O(l) \rightarrow CH_3CHO(l)$$

Use enthapies of combustion for acetylene and acetaldehyde to compute the enthalpy of the above reaction.

$$C_2H_2(g) + 2O_2(g) \rightarrow 2CO_2(g) + H_2O(l)$$

$\Delta H_c = -1299.6$ kJ/mol

$$CH_3CHO(l) + 2O_2(g) \rightarrow 2CO_2(g) + 2H_2O(l)$$

$\Delta H_c = -1166.9$ kJ/mol

**12.** Calculate the enthalpy for the combustion of decane. $\Delta H_f^0$ for liquid decane is $-300.9$ kJ/mol.

$$C_{10}H_{22}(l) + 15O_2(g) \rightarrow 10CO_2(g) + 11H_2O(l)$$

**13.** Find the enthalpy of the reaction of magnesium oxide with hydrogen chloride.

$$MgO(s) + 2HCl(g) \rightarrow MgCl_2(s) + H_2O(l)$$

Use the following equations and data.

| | |
|---|---|
| $Mg(s) + 2HCl(g) \rightarrow MgCl_2(s) + H_2(g)$ | $\Delta H = -456.9$ kJ/mol |
| $Mg(s) + O_2(g) \rightarrow MgO(s)$ | $\Delta H = -601.6$ kJ/mol |
| $H_2O(l) \rightarrow H_2(g) + O_2(g)$ | $\Delta H = +285.8$ kJ/mol |

---

**Problem Solving** *continued*

**14.** What is the Gibbs energy change for the following reaction at 25°C?

$$2NaOH(s) + 2Na(s) \xrightarrow{\Delta} 2Na_2O(s) + H_2(g)$$

$\Delta S = 10.6$ J/mol·K

$\Delta H^0_{f_{NaOH}} = -425.9$ kJ/mol

**15.** The following equation represents the reaction between gaseous HCl and gaseous ammonia to form solid ammonium chloride.

$$NH_3(g) + HCl(g) \rightarrow NH_4Cl(s)$$

Calculate the entropy change in J/mol·K for the reaction of hydrogen chloride and ammonia at 25°C using the following data and the values fround in **Table 1.**

$\Delta G = -91.2$ kJ/mol

**16.** The production of steel from iron involves the removal of many impurities in the iron ore. The following equations show some of the purifying reactions. Calculate the enthalpy for each reaction. Use **Table 1** and the data given.

**a.** $3C(s) + Fe_2O_3(s) \rightarrow 3CO(g) + 2Fe(s)$

$\Delta H^0_{f_{CO(g)}} = -110.53$ kJ/mol

**b.** $3Mn(s) + Fe_2O_3(s) \rightarrow 3MnO(s) + 2Fe(s)$

$\Delta H^0_{f_{MnO(s)}} = -384.9$ kJ/mol

**c.** $12P(s) + 10Fe_2O_3(s) \rightarrow 3P_4O_{10}(s) + 20Fe(s)$

$\Delta H^0_{f_{P_4O_{10}(s)}} = -3009.9$ kJ/mol

**d.** $3Si(s) + 2Fe_2O_3(s) \rightarrow 3SiO_2(s) + 4Fe(s)$

$\Delta H^0_{f_{SiO_2(s)}} = -910.9$ kJ/mol

**e.** $3S(s) + 2Fe_2O_3(s) \rightarrow 3SO_2(g) + 4Fe(s)$

Name _____ Class _____ Date _____

# Problem Solving

## Gas Laws

Chemists found that there were relationships among temperature, volume, pressure, and quantity of a gas that could be described mathematically. This chapter deals with Boyle's law, Charles's law, Gay-Lussac's law, the combined gas law, and Dalton's law of partial pressures. These laws have one condition in common. They all assume that the molar amount of gas does not change. In other words, these laws work correctly only when no additional gas enters a system and when no gas leaks out of it. Remember also that a law describes a fact of nature. Gases do not "obey" laws. The law does not dictate the behavior of the gas. Rather, each gas law describes a certain behavior of gas that occurs if conditions are right.

### BOYLE'S LAW

Robert Boyle, a British chemist who lived from 1627 to 1691 formulated the first gas law, now known as Boyle's law. This law describes the relationship between the pressure and volume of a sample of gas confined in a container. Boyle found that gases compress, much like a spring, when the pressure on the gas is increased. He also found that they "spring back" when the pressure is lowered. By "springing back" he meant that the volume increases when pressure is lowered. It's important to note that Boyle's law is true only if the temperature of the gas does not change and no additional gas is added to the container or leaks out of the container.

Boyle's law states that *the volume and pressure of a sample of gas are* <u>*inversely*</u> *proportional to each other at constant temperature*. This statement can be expressed as follows.

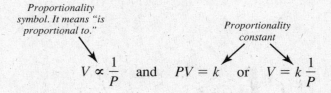

$$V \propto \frac{1}{P} \quad \text{and} \quad PV = k \quad \text{or} \quad V = k\,\frac{1}{P}$$

According to Boyle's law, when the pressure on a gas is *increased*, the volume of the gas *decreases*. For example, if the pressure is doubled, the volume decreases by half. If the volume quadruples, the pressure decreases to one-fourth of its original value.

The expression $PV = k$ means that the product of the pressure and volume of any sample of gas is a constant, $k$. If this is true, then $P \times V$ under one set of conditions is equal to $P \times V$ for the same sample of gas under a second set of conditions, as long as the temperature remains constant.

**Problem Solving** *continued*

Boyle's law can be expressed by the following mathematical equation.

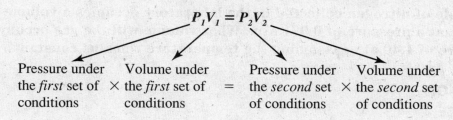

$$P_1V_1 = P_2V_2$$

| Pressure under the *first* set of conditions | × | Volume under the *first* set of conditions | = | Pressure under the *second* set of conditions | × | Volume under the *second* set of conditions |

**General Plan for Solving Boyle's-Law Problems**

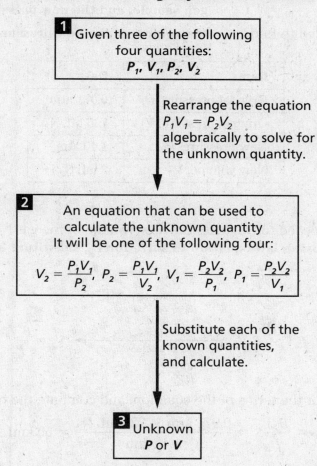

**1** Given three of the following four quantities:
$P_1$, $V_1$, $P_2$, $V_2$

Rearrange the equation
$P_1V_1 = P_2V_2$
algebraically to solve for
the unknown quantity.

**2** An equation that can be used to calculate the unknown quantity It will be one of the following four:

$$V_2 = \frac{P_1V_1}{P_2}, \quad P_2 = \frac{P_1V_1}{V_2}, \quad V_1 = \frac{P_2V_2}{P_1}, \quad P_1 = \frac{P_2V_2}{V_1}$$

Substitute each of the known quantities, and calculate.

**3** Unknown
$P$ or $V$

**| Problem Solving** *continued*

## Sample Problem 1

A sample of nitrogen collected in the laboratory occupies a volume of 725 mL at a pressure of 0.971 atm. What volume will the gas occupy at a pressure of 1.40 atm, assuming the temperature remains constant?

## Solution
### ANALYZE

What is given in the problem?     the original volume and pressure of the nitrogen sample, and the new pressure of the sample

What are you asked to find?     the volume at the new pressure

| Items | Data |
|---|---|
| Original pressure, $P_1$ | 0.971 atm |
| Original pressure, $V_1$ | 725 mL $N_2$ |
| New pressure, $P_2$ | 1.40 atm |
| New volume, $V_2$ | ? mL $N_2$ |

### PLAN

What steps are needed to calculate the new volume of the gas?
**Rearrange the Boyle's law equation to solve for $V_2$, substitute known quantities, and calculate.**

$$\boxed{1} \quad P_1V_1 = P_2V_2 \xrightarrow[\text{insert data and solve for } V_2]{\substack{\text{to solve for } V_2, \\ \text{divide both sides of} \\ \text{the equation by } P_2 \text{ to} \\ \text{isolate } V_2}} V_2 = \frac{P_1V_1}{P_2} \quad \boxed{2}$$

### COMPUTE
Substitute data for the terms of the equation, and compute the result.

$$V_2 = \frac{P_1V_1}{P_2} = \frac{0.971 \text{ atm} \times 725 \text{ mL } N_2}{1.40 \text{ atm}} = 503 \text{ mL } N_2$$

### EVALUATE
Are the units correct?
**Yes; units canceled to give mL $N_2$.**

Is the number of significant figures correct?
**Yes; the number of significant figures is correct because data were given to three significant figures.**

Is the answer reasonable?
**Yes; pressure increased by about 1/3, volume must decrease by about 1/3.**

**▌Problem Solving** *continued*

# Practice

**In each of the following problems, assume that the temperature and molar quantity of gas do not change.**

1. Calculate the unknown quantity in each of the following measurements of gases.

| | $P_1$ | $V_1$ | $P_2$ | $V_2$ | |
|---|---|---|---|---|---|
| **a.** | 3.0 atm | 25 mL | 6.0 atm | ? mL | ans: 13 mL |
| **b.** | 99.97 kPa | 550. mL | ? kPa | 275 mL | ans: 200. kPa |
| **c.** | 0.89 atm | ? L | 3.56 atm | 20.0 L | ans: 80. L |
| **d.** | ? kPa | 800. mL | 500. kPa | 160. mL | ans: 100. kPa |
| **e.** | 0.040 atm | ? L | 250 atm | $1.0 \times 10^{-2}$ L | ans: 63 L |

**| Problem Solving** *continued*

**2.** A sample of neon gas occupies a volume of 2.8 L at 1.8 atm. What will its volume be at 1.2 atm? **ans: 4.2 L**

**3.** To what pressure would you have to compress 48.0 L of oxygen gas at 99.3 kPa in order to reduce its volume to 16.0 L? **ans: 298 kPa**

**4.** A chemist collects 59.0 mL of sulfur dioxide gas on a day when the atmospheric pressure is 0.989 atm. On the next day, the pressure has changed to 0.967 atm. What will the volume of the $SO_2$ gas be on the second day? **ans: 60.3 mL**

**5.** 2.2 L of hydrogen at 6.5 atm pressure is used to fill a balloon at a final pressure of 1.15 atm. What is its final volume? **ans: 12 L**

## CHARLES'S LAW

The French physicist Jacques Charles carried out experiments in 1786 and 1787 that showed a relationship between the temperature and volume of gases at constant pressure. You know that most matter expands as its temperature rises. Gases are no different. When Benjamin Thomson and Lord Kelvin proposed an absolute temperature scale in 1848, it was possible to set up the mathematical expression of Charles's law.

**❚ Problem Solving** *continued*

Charles's law states that *the volume of a sample of gas is <u>directly</u> proportional to the absolute temperature when pressure remains constant.* Charles's law can be expressed as follows.

$$V \propto T \quad \text{and} \quad \frac{V}{T} = k, \quad \text{or} \quad V = kT$$

According to Charles's law, when the temperature of a sample of gas *increases*, the volume of the gas *increases* by the same factor. Therefore, doubling the Kelvin temperature of a gas will double its volume. Reducing the Kelvin temperature by 25% will reduce the volume by 25%.

The expression $V/T = k$ means that the result of volume divided by temperature is a constant, $k$, for any sample of gas. If this is true, then $V/T$ under one set of conditions is equal to $V/T$ for the same sample of gas under another set of conditions, as long as the pressure remains constant.

Charles's law can be expressed by the following mathematical equation.

$$\frac{V_1}{T_1} = \frac{V_2}{T_2}$$

### General Plan for Solving Charles's-Law Problems

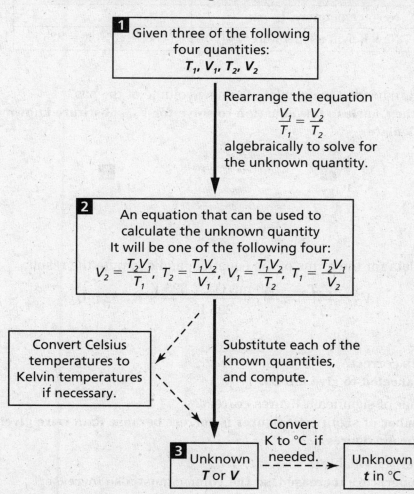

## Sample Problem 2

**A container of oxygen has a volume of 349 mL at a temperature of 22°C. What volume will the gas occupy at 50.°C?**

## Solution

### ANALYZE

What is given in the problem?    **the original volume and temperature of the oxygen sample, and the new temperature of the sample**

What are you asked to find?    **the volume at the new temperature**

| Items | Data |
|---|---|
| Original volume, $V_1$ | 349 mL $O_2$ |
| Original temperature, $t_1$ | 22°C |
| Original Kelvin temperature, $T_1$ | (22 + 273) K = 295 K |
| New volume, $V_2$ | ? mL $O_2$ |
| New temperature, $t_2$ | 50.°C |
| New Kelvin temperature, $T_2$ | ? (50. + 273) K = 323 K |

### PLAN

What steps are needed to calculate the new volume of the gas?
**Rearrange the Charles's law equation to solve for $V_2$, substitute known quantities, and calculate.**

**1**
$$\frac{V_1}{T_1} = \frac{V_2}{T_2}$$

*to solve for $V_2$, multiply both sides of the equation by $T_2$ to isolate $V_2$*
*insert data and solve for $V_2$*

**2**
$$\frac{V_1 T_2}{T_1} = V_2$$

### COMPUTE

Substitute data for the terms of the equation, and compute the result.

$$V_2 = \frac{V_1 T_2}{T_1} = \frac{349 \text{ mL } O_2 \times 323 \text{ K}}{295 \text{ K}} = 382 \text{ mL } O_2$$

### EVALUATE

Are the units correct?
**Yes; units canceled to give mL $O_2$.**

Is the number of significant figures correct?
**Yes; the number of significant figures is correct because data were given to three significant figures.**

Is the answer reasonable?
**Yes; the temperature increased, so the volume must also increase.**

**Problem Solving** *continued*

# Practice

**In each of the following problems, assume that the pressure and molar quantity of gas do not change.**

1. Calculate the unknown quantity in each of the following measurements of gases.

|     | $V_1$ | $T_1$ | $V_2$ | $T_2$ |  |
| --- | --- | --- | --- | --- | --- |
| **a.** | 40.0 mL | 280. K | ? mL | 350. K | **ans: 50.0 mL** |
| **b.** | 0.606 L | 300. K | 0.404 L | ? K | **ans: 200. K** |
| **c.** | ? mL | 292 K | 250. mL | 365 K | **ans: 200. mL** |
| **d.** | 100. mL | ? K | 125 mL | 305 K | **ans: 244 K** |
| **e.** | 0.0024 L | 22°C | ? L | −14°C | **ans: 0.0021 L** |

**▌Problem Solving** *continued*

**2.** A balloon full of air has a volume of 2.75 L at a temperature of 18°C. What is the balloon's volume at 45°C? **ans: 3.01 L**

**3.** A sample of argon has a volume of 0.43 mL at 24°C. At what temperature in degrees Celsius will it have a volume of 0.57 mL? **ans: 121°C**

## GAY-LUSSAC'S LAW

You may have noticed a warning on an aerosol spray can that says something similar to *Do not incinerate! Do not expose to temperatures greater than 140°F!* Warnings such as this appear because the pressure of a confined gas increases with increasing temperature. If the temperature of the can increases enough, the can will explode because of the pressure that builds up inside of it.

The relationship between the pressure and temperature of a gas is described by Gay-Lussac's law. Gay-Lussac's law states that *the pressure of a sample of gas is __directly__ proportional to the absolute temperature when volume remains constant.* Gay-Lussac's law can be expressed as follows.

$$P \propto T \quad \text{and} \quad \frac{P}{T} = k, \quad \text{or} \quad P = kT$$

According to Gay-Lussac's law, when the temperature of a sample of gas *increases*, the pressure of the gas *increases* by the same factor. Therefore, doubling the temperature of a gas will double its pressure. Reducing the temperature of a gas to 75% of its original value will also reduce the pressure to 75% of its original value.

The expression $P/T = k$ means that the result of pressure divided by temperature is a constant, $k$, for any sample of gas. If this is true, then $P/T$ under one set of conditions is equal to $P/T$ for the same sample of gas under another set of conditions, as long as the volume remains constant.

Gay-Lussac's law can be expressed by the following mathematical equation.

$$\frac{P_1}{T_1} = \frac{P_2}{T_2}$$

**General Plan for Solving Gay-Lussac's-Law Problems**

```
┌─────────────────────────────┐
│ 1  Given three of the        │
│    following four quantities:│
│    T₁, P₁, T₂, P₂            │
└─────────────────────────────┘
```

$T_1$, $P_1$, $T_2$, $P_2$

Rearrange the equation
$$\frac{P_1}{T_1} = \frac{P_2}{T_2}$$
algebraically to solve for the unknown quantity.

```
┌─────────────────────────────────────┐
│ 2   An equation that can be used to  │
│     calculate the unknown quantity   │
│     It will be one of the following  │
│     four:                            │
└─────────────────────────────────────┘
```

$$P_2 = \frac{T_2 P_1}{T_1}, \quad T_2 = \frac{T_1 P_2}{P_1}, \quad P_1 = \frac{T_1 P_2}{T_2}, \quad T_1 = \frac{T_2 P_1}{P_2}$$

Convert Celsius temperatures to Kelvin temperatures if necessary.

Substitute each of the known quantities, and compute.

Convert K to °C if needed.

```
┌──────────────┐           ┌──────────────┐
│ 3  Unknown   │  ──────→  │   Unknown    │
│    T or P    │           │   t in °C    │
└──────────────┘           └──────────────┘
```

**| Problem Solving** *continued*

## Sample Problem 3

**A cylinder of gas has a pressure of 4.40 atm at 25°C. At what temperature in °C will it reach a pressure of 6.50 atm?**

## Solution

### ANALYZE

What is given in the problem?     **the original pressure and temperature of the gas in the cylinder, and the new pressure of the sample**

What are you asked to find?     **the temperature at which the gas reaches the specified pressure**

| Items | Data |
|---|---|
| Original pressure, $P_1$ | 4.40 atm |
| Original temperature, $t_1$ | 25°C |
| Original Kelvin temperature, $T_1$ | (25 + 273) K = 298 K |
| New pressure, $P_2$ | 6.50 atm |
| New temperature, $t_2$ | ?°C |
| New Kelvin temperature, $T_2$ | ? K |

### PLAN

What steps are needed to calculate the new temperature of the gas?
**Rearrange the Gay-Lussac's law equation to solve for $T_2$, substitute known quantities, and calculate.**

$$\boxed{1} \quad \frac{P_1}{T_1} = \frac{P_2}{T_2} \quad \xrightarrow[\substack{\text{insert data and solve for } T_2}]{\substack{\textit{to solve for } T_2, \textit{divide} \\ \textit{both sides of the equation} \\ \textit{by } P_2 \textit{ and invert the} \\ \textit{equality to isolate } T_2}} \quad \boxed{2} \quad \frac{T_1 P_2}{P_1} = T_2$$

### COMPUTE

Substitute data for the terms of the equation, and compute the result.

$$T_2 = \frac{T_1 P_2}{P_1} = \frac{298 \text{ K} \times 6.50 \text{ atm}}{4.40 \text{ atm}} = 440. \text{ K}$$

Convert back to °C.

$$440. \text{ K} = (440. - 273)°C = 167°C$$

### EVALUATE

Are the units correct?
**Yes; units canceled to give Kelvin temperature, which was converted to °C.**

**| Problem Solving** *continued*

Is the number of significant figures correct?

**Yes; the number of significant figures is correct because the data were given to three significant figures.**

Is the answer reasonable?

**Yes; the pressure increases, so the temperature must also increase.**

# Practice

**In each of the following problems, assume that the volume and molar quantity of gas do not change.**

1. Calculate the unknown quantity in each of the following measurements of gases:

|     | $P_1$ | $T_1$ | $P_2$ | $T_2$ |                    |
| --- | ----- | ----- | ----- | ----- | ------------------ |
| **a.** | 1.50 atm | 273 K | ? atm | 410 K | **ans: 2.25 atm** |
| **b.** | 0.208 atm | 300. K | 0.156 atm | ? K | **ans: 225 K** |
| **c.** | ? kPa | 52°C | 99.7 kPa | 77°C | **ans: 92.6 kPa** |
| **d.** | 5.20 atm | ?°C | 4.16 atm | −13°C | **ans: 52°C** |
| **e.** | $8.33 \times 10^{-4}$ atm | −84°C | $3.92 \times 10^{-3}$ atm | ?°C | **ans: 616°C** |

**▌Problem Solving** *continued*

**2.** A cylinder of compressed gas has a pressure of 4.882 atm on one day. The next day, the same cylinder of gas has a pressure of 4.690 atm, and its temperature is 8°C. What was the temperature on the previous day in °C? **ans: 20.°C**

**3.** A mylar balloon is filled with helium gas to a pressure of 107 kPa when the temperature is 22°C. If the temperature changes to 45°C, what will be the pressure of the helium in the balloon? **ans: 115 kPa**

## THE COMBINED GAS LAW

Look at the relationships among temperature, volume, and pressure of a gas that you have studied so far.

| **Boyle's law** At constant temperature: $PV = k$ | **Charles's law** At constant pressure: $\dfrac{V}{T} = k$ | **Gay-Lussac's law** At constant volume: $\dfrac{P}{T} = k$ |
|---|---|---|

Notice in these proportions that while $P$ and $V$ are inversely proportional to each other, they are each directly proportional to temperature. These three gas laws can be combined in one *combined gas law*. This law can be expressed as follows.

$$\frac{PV}{T} = k$$

If $PV/T$ equals a constant, $k$, then $PV/T$ for a sample of gas under one set of conditions equals $PV/T$ under another set of conditions, assuming the amount of gas remains the same.

Therefore, the combined gas law can be expressed by the following mathematical equation. This equation can be used to solve problems in which pressure,

| **Problem Solving** *continued*

volume, and temperature of a gas vary. Only the molar quantity of the gas must be constant.

$$\frac{P_1 V_1}{T_1} = \frac{P_2 V_2}{T_2}$$

### General Plan for Solving Combined-Gas-Law Problems

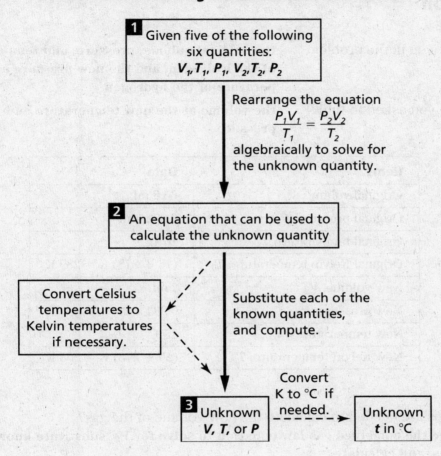

**1** Given five of the following six quantities: $V_1, T_1, P_1, V_2, T_2, P_2$

Rearrange the equation
$$\frac{P_1 V_1}{T_1} = \frac{P_2 V_2}{T_2}$$
algebraically to solve for the unknown quantity.

**2** An equation that can be used to calculate the unknown quantity

Convert Celsius temperatures to Kelvin temperatures if necessary.

Substitute each of the known quantities, and compute.

Convert K to °C if needed.

**3** Unknown $V$, $T$, or $P$

Unknown $t$ in °C

**Problem Solving** *continued*

## Sample Problem 4

**A sample of hydrogen gas has a volume of 65.0 mL at a pressure of 0.992 atm and a temperature of 16°C. What volume will the hydrogen occupy at 0.984 atm and 25°C?**

## Solution
### ANALYZE

What is given in the problem?    the original volume, pressure, and temperature of the hydrogen, and the new pressure and temperature of the hydrogen

What are you asked to find?    the volume at the new temperature and pressure

| Items | Data |
|---|---|
| Original volume, $V_1$ | 65.0 mL |
| Original pressure, $P_1$ | 0.992 atm |
| Original temperature, $t_1$ | 16°C |
| Original Kelvin temperature, $T_1$ | (16 + 273) K = 289 K |
| New volume, $V_2$ | ? mL |
| New pressure, $P_2$ | 0.984 atm |
| New temperature, $t_2$ | 25°C |
| New Kelvin temperature, $T_2$ | (25 + 273) K = 298 K |

### PLAN
What steps are needed to calculate the new volume of the gas?
**Rearrange the combined gas law equation to solve for $V_2$, substitute known quantities, and calculate.**

$$\boxed{1}\quad \frac{P_1V_1}{T_1} = \frac{P_2V_2}{T_2} \xrightarrow[\substack{\text{insert data and solve for } V_2}]{\substack{\text{to solve for } V_2, \\ \text{multiply both sides of} \\ \text{the equation by } T_2/P_2 \text{ to} \\ \text{isolate } V_2}} \boxed{2}\quad \frac{T_2P_1V_1}{P_2T_1} = V_2$$

### COMPUTE

Substitute data for the terms of the equation, and compute the result.

$$V_2 = \frac{T_2P_1V_1}{P_2T_1} = \frac{298 \text{ K} \times 0.992 \text{ atm} \times 65.0 \text{ mL H}_2}{0.984 \text{ atm} \times 289 \text{ K}} = 67.6 \text{ mL H}_2$$

### EVALUATE
Are the units correct?
**Yes; units canceled to give mL of $H_2$.**

Is the number of significant figures correct?

**Yes; the number of significant figures is correct because data were given to three significant figures.**

Is the answer reasonable?

**Yes; the temperature increases, and the pressure decreases, both of which have the effect of making the volume larger.**

## Practice

**In each of the following problems, it is assumed that the molar quantity of gas does not change.**

1. Calculate the unknown quantity in each of the following measurements of gases.

|     | $P_1$      | $V_1$    | $T_1$ | $P_2$       | $V_2$   | $T_2$ |            |
|-----|------------|----------|-------|-------------|---------|-------|------------|
| **a.** | 99.3 kPa   | 225 mL   | 15°C  | 102.8 kPa   | ? mL    | 24°C  | ans: 224 mL |
| **b.** | 0.959 atm  | 3.50 L   | 45°C  | ? atm       | 3.70 L  | 37°C  | ans: 0.884 atm |
| **c.** | 0.0036 atm | 62 mL    | 373 K | 0.0029 atm  | 64 mL   | ? K   | ans: 310 K |
| **d.** | 100. kPa   | 43.2 mL  | 19°C  | 101.3 kPa   | ? mL    | 0°C   | ans: 39.9 mL |

2. A student collects 450. mL of HCl(*g*) hydrogen chloride gas at a pressure of 100. kPa and a temperature of 17°C. What is the volume of the HCl at 0°C and 101.3 kPa? **ans: 418 mL**

| Problem Solving *continued*

## DALTON'S LAW OF PARTIAL PRESSURES

Air is a mixture of approximately 78% $N_2$, 20% $O_2$, 1% Ar, and 1% other gases by volume, so at any barometric pressure 78% of that pressure is exerted by nitrogen, 20% by oxygen, and so on. This phenomenon is described by Dalton's law of partial pressures, which says that *the total pressure of a mixture of gases is equal to the sum of the partial pressures of the component gases.* It can be stated mathematically as follows.

$$P_{Total} = P_{Gas\ 1} + P_{Gas\ 2} + P_{Gas\ 3} + P_{Gas\ 4} + \ldots$$

A common method of collecting gas samples in the laboratory is to bubble the gas into a bottle filled with water and allow it to displace the water. When this technique is used, however, the gas collected in the bottle contains a small but significant amount of water vapor. As a result, the pressure of the gas that has displaced the liquid water is the sum of the pressure of the gas plus the vapor pressure of water at that temperature. The vapor pressures of water at various temperatures are given in **Table 1** below.

## TABLE 1 VAPOR PRESSURE OF WATER

| Temp in °C | Vapor pressure in kPa | Temp in °C | Vapor pressure in kPa | Temp in °C | Vapor pressure in kPa |
|---|---|---|---|---|---|
| 10 | 1.23 | 17 | 1.94 | 24 | 2.98 |
| 11 | 1.31 | 18 | 2.06 | 25 | 3.17 |
| 12 | 1.40 | 19 | 2.19 | 26 | 3.36 |
| 13 | 1.50 | 20 | 2.34 | 27 | 3.57 |
| 14 | 1.60 | 21 | 2.49 | 28 | 3.78 |
| 15 | 1.71 | 22 | 2.64 | 29 | 4.01 |
| 16 | 1.82 | 23 | 2.81 | 30 | 4.25 |

$$P_{Total} = P_{Gas} + P_{H_2O\ vapor}$$

To find the true pressure of the gas alone, the pressure of the water vapor must be subtracted from the total pressure.

$$P_{Gas} = P_{Total} - P_{H_2O\ vapor}$$

You can use this corrected pressure in gas-law calculations to determine what the volume of the gas alone would be.

| Problem Solving *continued*

## Sample Problem 5

A student collects oxygen gas by water displacement at a temperature of 16°C. The total volume is 188 mL at a pressure of 92.3 kPa. What is the pressure of oxygen collected?

## Solution

### ANALYZE

What is given in the problem?    the total pressure, the fact that the gas was collected by water displacement, and the temperature

What are you asked to find?    the pressure of oxygen collected

| Items | Data |
|---|---|
| Total pressure, $P_{total}$ | 92.3 kPa |
| Temperature | 16°C |
| Vapor pressure of water at 16°C, $P_{H_2O}$* | 1.82 kPa |
| Pressure of collected $O_2$, $P_{O_2}$ | ? kPa |

*determined from Table 1

### PLAN

What step is needed to calculate the pressure of the oxygen collected?
**Use Dalton's law of partial pressures to determine the pressure of the oxygen alone in the container.**

Use the following equation to determine the pressure of oxygen alone.

$$P_{O_2} = \overset{given}{P_{total}} - \overset{from\ Table\ 1}{P_{H_2O\ vapor}}$$

### COMPUTE

$$P_{O_2} = P_{total} - P_{H_2O\ vapor} = 92.3\ kPa - 1.82\ kPa = 90.5\ kPa$$

### EVALUATE

Are the units correct?
**Yes; the units should be kPa.**

Is the number of significant figures correct?
**Yes; the number of significant figures is correct because the total pressure was given to one decimal place.**

Is the answer reasonable?
**Yes; the pressure of the collected oxygen and the water vapor add up to the total pressure of the system.**

## Practice

**1.** A chemist collects a sample of $H_2S(g)$ over water at a temperature of 27°C. The total pressure of the gas that has displaced a volume of 15 mL of water is 207.33 kPa. What is the pressure of the $H_2S$ gas collected? **ans: 203.76 kPa**

# Sample Problem 6

Chlorine gas is collected by water displacement at a temperature of 19°C. The total volume is 1.45 L at a pressure of 156.5 kPa. Assume that no chlorine gas dissolves in the water. What is the volume of chlorine corrected to STP?

# Solution

## ANALYZE

What is given in the problem?    the total initial volume, pressure, and temperature; the fact that the gas was collected by water displacement; and the final temperature and pressure of the chlorine

What are you asked to find?    the volume of chlorine at STP

| Items | Data |
|---|---|
| Total original volume, $V_1$ | 1.45 L |
| Total original pressure, $P_{total}$ | 156.5 kPa |
| Original temperature, $t_1$ | 19°C |
| Original Kelvin temperature, $T_1$ | (19 + 273) K = 292 K |
| Vapor pressure of water at 19°C, $P_{H_2O}$ | 2.19 kPa |
| Pressure of collected $Cl_2$, $P_1$ | ? kPa |
| New volume, $V_2$ | ? mL |
| New pressure, $P_2$ | 101.3 kPa |
| New temperature, $t_2$ | 0°C |
| New Kelvin temperature, $T_2$ | 273 K |

## PLAN

What steps are needed to calculate the volume of the chlorine at STP?

Use Dalton's law of partial pressures to determine the pressure of the chlorine alone in the container. Use the combined gas law to solve for the new volume.

$$P_1 = \overset{given}{P_{total}} - \overset{from\ table}{P_{H_2O\ vapor}}$$

Calculate the volume of chlorine at STP using the following equation.

$$\frac{P_1V_1}{T_1} = \frac{P_2V_2}{T_2}$$

Solve for $V_2$.

$$\frac{\overset{calculated\ above}{P_1}\ \overset{given}{T_2V_1}}{\underset{given}{P_2T_1}} = V_2$$

**| Problem Solving** *continued*

## COMPUTE
Determine the pressure of chlorine alone.

$$P_1 = P_{total} - P_{H_2O\ vapor} = 156.5\ \text{kPa} - 2.19\ \text{kPa} = 154.3\ \text{kPa}$$

Calculate the volume of $Cl_2$ at STP.

$$V_2 = \frac{154.3\ \text{kPa} \times 273\ \text{K} \times 1.45\ \text{L}}{101.3\ \text{kPa} \times 292\ \text{K}} = 2.06\ \text{L}$$

## EVALUATE
Are the units correct?
**Yes; the units canceled leaving the correct units, L.**

Is the number of significant figures correct?
**Yes; the number of significant figures is correct because the data were given to a minimum of three significant figures.**

Is the answer reasonable?
**Yes; the pressure decreased, so the volume had to increase.**

## Practice

**In each of the following problems, assume that the molar quantity of gas does not change.**

1. Some hydrogen is collected over water at 10°C and 105.5 kPa pressure. The total volume of the sample was 1.93 L. Calculate the volume of the hydrogen corrected to STP. **ans: 1.92 L**

2. One student carries out a reaction that gives off methane gas and obtains a total volume by water displacement of 338 mL at a temperature of 19°C and a pressure of 0.9566 atm. Another student does the identical experiment on another day at a temperature of 26°C and a pressure of 0.989 atm. Which student collected more $CH_4$? **ans: The second student collected more $CH_4$.**

**| Problem Solving** *continued*

## Additional Problems

**In each of the following problems, assume that the molar quantity of gas does not change.**

1. Calculate the unknown quantity in each of the following measurements of gases.

|    | $P_1$ | $V_1$ | $P_2$ | $V_2$ |
|----|-------|-------|-------|-------|
| **a.** | 127.3 kPa | 796 cm$^3$ | ? kPa | 965 cm$^3$ |
| **b.** | $7.1 \times 10^2$ atm | ? mL | $9.6 \times 10^{-1}$ atm | $3.7 \times 10^3$ mL |
| **c.** | ? kPa | 1.77 L | 30.79 kPa | 2.44 L |
| **d.** | 114 kPa | 2.93 dm$^3$ | $4.93 \times 10^4$ kPa | ? dm$^3$ |
| **e.** | 1.00 atm | 120. mL | ? atm | 97.0 mL |
| **f.** | 0.77 atm | 3.6 m$^3$ | 1.90 atm | ? m$^3$ |

2. A gas cylinder contains 0.722 m$^3$ of hydrogen gas at a pressure of 10.6 atm. If the gas is used to fill a balloon at a pressure of 0.96 atm, what is the volume in m$^3$ of the filled balloon?

3. A weather balloon has a maximum volume of $7.50 \times 10^3$ L. The balloon contains 195 L of helium gas at a pressure of 0.993 atm. What will be the pressure when the balloon is at maximum volume?

4. A rubber ball contains $5.70 \times 10^{-1}$ dm$^3$ of gas at a pressure of 1.05 atm. What volume will the gas occupy at 7.47 atm?

5. Calculate the unknown quantity in each of the following measurements of gases.

|    | $V_1$ | $T_1$ | $V_2$ | $T_2$ |
|----|-------|-------|-------|-------|
| **a.** | 26.5 mL | ? K | 32.9 mL | 290. K |
| **b.** | ? dm$^3$ | 100.°C | 0.83 dm$^3$ | −9°C |
| **c.** | $7.44 \times 10^4$ mm$^3$ | 870.°C | $2.59 \times 10^2$ mm$^3$ | ?°C |
| **d.** | $5.63 \times 10^{-2}$ L | 132 K | ? L | 190. K |
| **e.** | ? cm$^3$ | 243 K | 819 cm$^3$ | 409 K |
| **f.** | 679 m$^3$ | −3°C | ? m$^3$ | −246°C |

6. A bubble of carbon dioxide gas in some unbaked bread dough has a volume of 1.15 cm$^3$ at a temperature of 22°C. What volume will the bubble have when the bread is baked and the bubble reaches a temperature of 99°C?

7. A perfectly elastic balloon contains 6.75 dm$^3$ of air at a temperature of 40.°C. What is the temperature if the balloon has a volume of 5.03 dm$^3$?

**8.** Calculate the unknown quantity in each of the following measurements of gases.

| | $P_1$ | $T_1$ | $P_2$ | $T_2$ |
|---|---|---|---|---|
| **a.** | 0.777 atm | ?°C | 5.6 atm | 192°C |
| **b.** | 152 kPa | 302 K | ? kPa | 11 K |
| **c.** | ? atm | −76°C | 3.97 atm | 27°C |
| **d.** | 395 atm | 46°C | 706 atm | ?°C |
| **e.** | ? atm | −37°C | 350. atm | 2050°C |
| **f.** | 0.39 atm | 263 K | 0.058 atm | ? K |

**9.** A 2 L bottle containing only air is sealed at a temperature of 22°C and a pressure of 0.982 atm. The bottle is placed in a freezer and allowed to cool to − 3°C. What is the pressure in the bottle?

**10.** The pressure in a car tire is 2.50 atm at a temperature of 33°C. What would the pressure be if the tire were allowed to cool to 0°C? Assume that the tire does not change volume.

**11.** A container filled with helium gas has a pressure of 127.5 kPa at a temperature of 290. K. What is the temperature when the pressure is 3.51 kPa?

**12.** Calculate the unknown quantity in each of the following measurements of gases.

| | $P_1$ | $V_1$ | $T_1$ | $P_2$ | $V_2$ | $T_2$ |
|---|---|---|---|---|---|---|
| **a.** | 1.03 atm | 1.65 L | 19°C | 0.920 atm | ? L | 46°C |
| **b.** | 107.0 kPa | 3.79 dm³ | 73°C | ? kPa | 7.58 dm³ | 217°C |
| **c.** | 0.029 atm | 249 mL | ? K | 0.098 atm | 197 mL | 293 K |
| **d.** | 113 kPa | ? mm³ | 12°C | 149 kPa | $3.18 \times 10^3$ mm³ | −18°C |
| **e.** | 1.15 atm | 0.93 m³ | −22°C | 1.01 atm | 0.85 m³ | ?°C |
| **f.** | ? atm | 156 cm³ | 195 K | 2.25 atm | 468 cm³ | 584 K |

**13.** A scientist has a sample of gas that was collected several days earlier. The sample has a volume of 392 cm³ at a pressure of 0.987 atm and a temperature of 21°C. On the day the gas was collected, the temperature was 13°C and the pressure was 0.992 atm. What volume did the gas have on the day it was collected?

**14.** Hydrogen gas is collected by water displacement. Total volume collected is 0.461 L at a temperature of 17°C and a pressure of 0.989 atm. What is the pressure of dry hydrogen gas collected?

15. One container with a volume of 1.00 L contains argon at a pressure of 1.77 atm, and a second container of 1.50 L volume contains argon at a pressure of 0.487 atm. They are then connected to each other so that the pressure can become equal in both containers. What is the equalized pressure? Hint: Each sample of gas now occupies the total space. Dalton's law of partial pressures applies here.

16. Oxygen gas is collected over water at a temperature of 10.°C and a pressure of 1.02 atm. The volume of gas plus water vapor collected is 293 mL. What volume of oxygen at STP was collected?

17. A 500 mL bottle is partially filled with water so that the total volume of gases (water vapor and air) remaining in the bottle is 325 $cm^3$, measured at 20.°C and 101.3 kPa. The bottle is sealed and taken to a mountaintop where the pressure is 76.24 kPa and the temperature is 10°C. If the bottle is upside down and the seal leaks, how much water will leak out? The key to this problem is to determine the pressure in the 325 $cm^3$ space when the bottle is at the top of the mountain.

18. An air thermometer can be constructed by using a glass bubble attached to a piece of small-diameter glass tubing. The tubing contains a small amount of colored water that rises when the temperature increases and the trapped air expands. You want a 0.20 $cm^3$ change in volume to equal a 1°C change in temperature. What total volume of air at 20.°C should be trapped in the apparatus below the liquid?

19. A sample of nitrogen gas is collected over water, yielding a total volume of 62.25 mL at a temperature of 22°C and a total pressure of 97.7 kPa. At what pressure will the nitrogen alone occupy a volume of 50.00 mL at the same temperature?

20. The theoretical yield of a reaction that gives off nitrogen trifluoride gas is 844 mL at STP. What total volume of $NF_3$ plus water vapor will be collected over water at 25°C and a total pressure of 1.017 atm?

21. A weather balloon is inflated with 2.94 kL of helium at a location where the pressure is 1.06 atm and the temperature is 32°C. What will be the volume of the balloon at an altitude where the pressure is 0.092 atm and the temperature is −35°C?

22. The safety limit for a certain can of aerosol spray is 95°C. If the pressure of the gas in the can is 2.96 atm when it is 17°C, what will the pressure be at the safety limit?

23. A chemistry student collects a sample of ammonia gas at a temperature of 39°C. Later, the student measures the volume of the ammonia as 108 mL, but its temperature is now 21°C. What was the volume of the ammonia when it was collected?

24. A quantity of $CO_2$ gas occupies a volume of 624 L at a pressure of 1.40 atm. If this $CO_2$ is pumped into a gas cylinder that has a volume of 80.0 L, what pressure will the $CO_2$ exert on the cylinder?

Skills Worksheet

# Problem Solving

## The Ideal Gas Law

In 1811, the Italian chemist Amedeo Avogadro proposed the principle that *equal volumes of gases at the same temperature and pressure contain equal numbers of molecules.* He determined that at standard temperature and pressure, one mole of gas occupies 22.414 10 L (usually rounded to 22.4 L).

At this point, if you know the number of moles of a gas, you can use the molar volume of 22.4 L/mol to calculate the volume that amount of gas would occupy at STP. Then you could use the combined gas law to determine the volume of the gas under any other set of conditions. However, a much simpler way to accomplish the same task is by using the ideal gas law.

The *ideal gas law* is a mathematical relationship that has the conditions of standard temperature (273 K) and pressure (1 atm or 101.3 kPa) plus the molar gas volume (22.4 L/mol) already combined into a single constant. The following equation is the mathematical statement of the ideal gas law.

$$PV = nRT$$

in which

$P$ = the pressure of a sample of gas

$V$ = the volume of a sample of gas

$n$ = the number of moles of gas present

$T$ = the Kelvin temperature of the gas

$R$ = the ideal gas constant, which combines standard conditions
and molar volume into a single constant

The value of the ideal gas constant, $R$, depends on the units of $P$ and $V$ being used in the equation. Temperature is always in kelvins and amount of gas is always in moles. The most common values used for $R$ are shown below.

| Units of $P$ and $V$ | Value of $R$ |
|---|---|
| Atmospheres and liters | $0.0821 \dfrac{L \cdot atm}{mol \cdot K}$ |
| Kilopascals and liters | $8.314 \dfrac{L \cdot kPa}{mol \cdot K}$ |

**Problem Solving** *continued*

If you have volume units other than liters or pressure units other than atmospheres or kilopascals, it is best to convert volume to liters and pressure to atmospheres or kilopascals.

**General Plan for Solving Ideal-Gas-Law Problems**

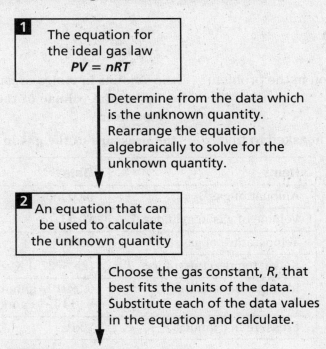

**1** The equation for the ideal gas law
$$PV = nRT$$

Determine from the data which is the unknown quantity. Rearrange the equation algebraically to solve for the unknown quantity.

**2** An equation that can be used to calculate the unknown quantity

Choose the gas constant, $R$, that best fits the units of the data. Substitute each of the data values in the equation and calculate.

**3** Unknown
$P, V, n,$ or $T$

## Sample Problem 1

An engineer pumps 5.00 mol of carbon monoxide gas into a cylinder that has a capacity of 20.0 L. What is the pressure in kPa of CO inside the cylinder at 25°C?

## Solution

### ANALYZE

What is given in the problem?     the amount in moles of gas pumped into the cylinder, the volume of the cylinder, and the temperature

What are you asked to find?     the pressure of the gas in the cylinder

| Items | Data |
|---|---|
| Amount of gas, $n$ | 5.00 mol |
| Volume of gas in cylinder, $V$ | 20.0 L |
| Temperature of gas, $t$ | 25°C |
| Kelvin temperature of gas, $T$ | (25 + 273) K = 298 K |
| Ideal gas constant, $R$ | 0.0821 L·atm/mol·K or 8.314 L·kPa/mol·K |
| Pressure in cylinder, $P$ | ? kPa |

### PLAN

What steps are needed to calculate the new pressure of the gas?

**Rearrange the ideal-gas-law equation to solve for $P$, substitute known quantities, and calculate.**

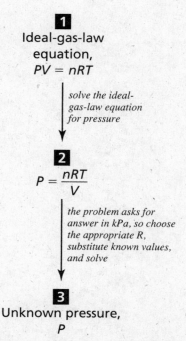

**1**
Ideal-gas-law
equation,
$PV = nRT$

*solve the ideal-gas-law equation for pressure*

**2**
$P = \dfrac{nRT}{V}$

*the problem asks for answer in kPa, so choose the appropriate R, substitute known values, and solve*

**3**
Unknown pressure,
$P$

**| Problem Solving** *continued*

$$PV = nRT$$

Solve the ideal-gas-law equation for $P$, the unknown quantity.

$$P = \frac{nRT}{V}$$

## COMPUTE

The problem asks for pressure in kPa, so use $R = 8.314$ L·kPa/mol·K.

$$P = \frac{5.00 \; \text{mol} \times 8.314 \; \text{L·kPa/mol·K} \times 298 \; \text{K}}{20.0 \; \text{L}} = 619 \; \text{kPa}$$

## EVALUATE

Are the units correct?

**Yes; the ideal gas constant was selected so that the units canceled to give kPa.**

Is the number of significant figures correct?

**Yes; the number of significant figures is correct because data were given to three significant figures.**

Is the answer reasonable?

**Yes; the calculation can be approximated as $(1/4) \times (8 \times 300)$, or 2400/4, which equals 600. Thus, 619 kPa is in the right range.**

# Practice

1. A student collects 425 mL of oxygen at a temperature of 24°C and a pressure of 0.899 atm. How many moles of oxygen did the student collect?

   ans: $1.57 \times 10^{-2}$ mol $O_2$

**Problem Solving** *continued*

**2.** Use the ideal-gas-law equation to calculate the unknown quantity in each of the following sets of measurements. You will need to convert Celsius temperatures to Kelvin temperatures and volume units to liters.

|    | P        | V        | n          | T        |                                    |
|----|----------|----------|------------|----------|------------------------------------|
| **a.** | 1.09 atm | ? L      | 0.0881 mol | 302 K    | **ans: 2.00 L**                    |
| **b.** | 94.9 kPa | 0.0350 L | ? mol      | 55°C     | **ans: $1.22 \times 10^{-3}$ mol** |
| **c.** | ? kPa    | 15.7 L   | 0.815 mol  | −20.°C   | **ans: 109 kPa**                   |
| **d.** | 0.500 atm | 629 mL  | 0.0337 mol | ? K      | **ans: 114 K**                     |
| **e.** | 0.950 atm | ? L     | 0.0818 mol | 19°C     | **ans: 2.06 L**                    |
| **f.** | 107 kPa  | 39.0 mL  | ? mol      | 27°C     | **ans: $1.67 \times 10^{-3}$ mol** |

**Problem Solving** *continued*

## APPLICATIONS OF THE IDEAL GAS LAW

You have seen that you can use the ideal gas law to calculate the moles of gas, $n$, in a sample when you know the pressure, volume, and temperature of the sample. When you know the amount and identity of the substance, you can use its molar mass to calculate its mass. You did this when you learned how to convert between mass and moles. The relationship is expressed as follows.

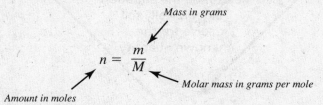

$$n = \frac{m}{M}$$

*Mass in grams*

*Amount in moles*

*Molar mass in grams per mole*

If you substitute the expression $m/M$ for $n$ in the ideal-gas-law equation, you get the following equation.

$$PV = \frac{m}{M}RT$$

This version of the ideal gas law can be solved for any of the five variables $P$, $V$, $m$, $M$, or $T$. It is especially useful in determining the molecular mass of a substance. This equation can also be related to the density of a gas. Density is mass per unit volume, as shown in the following equation.

$$D = \frac{m}{V}$$

Solve for $m$:

$$m = DV$$

Then, substitute $DV$ for $m$ in the gas law equation:

$$PV = \frac{DV}{M}RT$$

The two $V$ terms cancel and the equation is rearranged to give:

$$PM = DRT \quad \text{or} \quad D = \frac{PM}{RT}$$

This equation can be used to compute the density of a gas under any conditions of temperature and pressure. It can also be used to calculate the molar mass of an unknown gas if its density is known.

### General Plan for Solving Problems Involving
### Applications of the Ideal Gas Law

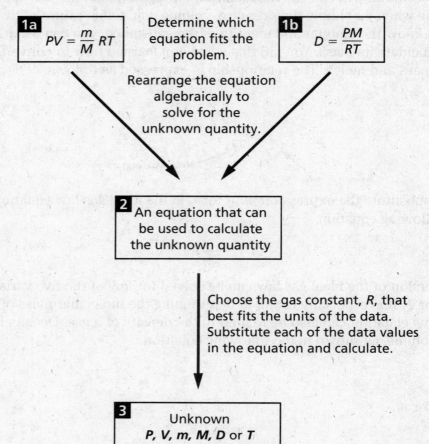

**1a**

$$PV = \frac{m}{M} RT$$

Determine which
equation fits the
problem.

**1b**

$$D = \frac{PM}{RT}$$

Rearrange the equation
algebraically to
solve for the
unknown quantity.

**2** An equation that can
be used to calculate
the unknown quantity

Choose the gas constant, *R*, that
best fits the units of the data.
Substitute each of the data values
in the equation and calculate.

**3** Unknown
***P, V, m, M, D*** or ***T***

# Sample Problem 2

Determine the molar mass of an unknown gas that has a volume of 72.5 mL at a temperature of 68°C, a pressure of 0.980 atm, and a mass of 0.207 g.

# Solution

## ANALYZE

What is given in the problem?    **the mass, pressure, volume, and temperature of the gas**

What are you asked to find?    **the molar mass of the gas**

| Items | Data |
|---|---|
| Volume of gas, $V$ | 72.5 mL |
| Temperature of gas, $t$ | 68°C |
| Kelvin temperature of gas, $T$ | (68 + 273) K = 341 K |
| Pressure of gas, $P$ | 0.980 atm |
| Mass of gas, $m$ | 0.207 g |
| Ideal gas constant, $R$ | 8.314 L·kPa/mol·K or 0.0821 L·atm/mol·K |
| Molar mass of gas, $M$ | ? g/mol |

## PLAN

What steps are needed to calculate the new volume of the gas?

**Select the equation that will give the desired result. Solve the equation for the unknown quantity. Substitute data values into the solved equation, and calculate.**

**1a**

$$PV = \frac{m}{M} RT$$

*solve this equation for molar mass*

**2**

$$M = \frac{mRT}{PV}$$

*the problem gives pressure in atm, so choose the appropriate R, substitute known values, and solve*

**3**

Unknown molar mass, $M$

**❚ Problem Solving** *continued*

Use the equation that includes $m$ and $M$.

$$PV = \frac{m}{M}RT$$

Solve the equation for $M$, the unknown quantity.

$$M = \frac{mRT}{PV}$$

**COMPUTE**

Convert the volume in milliliters to liters

$$72.5 \text{ mL} \times \frac{1 \text{ L}}{1000 \text{ mL}} = 0.0725 \text{ L}$$

The data give pressure in atm, so use $R = 0.0821$ L·atm/mol·K.

$$M = \frac{mRT}{PV} = \frac{0.207 \text{ g} \times 0.0821 \text{ L·atm/mol·K} \times 341 \text{ K}}{0.980 \text{ atm} \times 0.0725 \text{ L}} = 81.6 \text{ g/mol}$$

**EVALUATE**

Are the units correct?

**Yes; units canceled to give g/mol, the correct units for molar mass.**

Is the number of significant figures correct?

**Yes; the number of significant figures is correct because data were given to three significant figures.**

Is the answer reasonable?

**Yes; 81.6 g/mol is a reasonable molar mass. The calculation can be approximated as 0.2 × 341 × (8/7), which is roughly 80.**

## Practice

1. A sample of an unknown gas has a mass of 0.116 g. It occupies a volume of 25.0 mL at a temperature of 127°C and has a pressure of 155.3 kPa. Calculate the molar mass of the gas. **ans: 99.4 g/mol**

2. Determine the mass of $CO_2$ gas that has a volume of 7.10 L at a pressure of 1.11 atm and a temperature of 31°C. Hint: Solve the equation for $m$, and calculate the molar mass using the chemical formula and the periodic table. **ans: 13.9 g**

# Sample Problem 3

**Determine the density of hydrogen bromide gas at 3.10 atm and −5°C.**

# Solution

## ANALYZE

What is given in the problem?     **the pressure and temperature of the HBr gas**

What are you asked to find?       **the density of the gas**

| Items | Data |
|---|---|
| Temperature of HBr, $t$ | −5°C |
| Kelvin temperature of HBr, $T$ | (−5 + 273) K = 268 K |
| Pressure of HBr, $P$ | 3.10 atm |
| Molar mass of HBr, $M^*$ | 80.91 g/mol |
| Ideal gas constant, $R$ | 8.314 L·kPa/mol·K or 0.0821 L·atm/mol·K |
| Density of HBr, $D$ | ? g/L |

*determined from the periodic table

## PLAN

What steps are needed to calculate the density of HBr under the conditions given?

**Select the equation that will give the desired result. Rearrange the equation to solve for the unknown quantity. Substitute data values into the correct equation, and calculate.**

**1**

$$D = \frac{PM}{RT}$$

*equation is already written correctly to solve for the unknown*

**2**

$$D = \frac{PM}{RT}$$

*the problem gives pressure in atm, so choose the appropriate R, substitute known values, and solve*

**3**

Unknown density, $D$

Use the equation that includes density.

$$D = \frac{PM}{RT}$$

**COMPUTE**

The data give pressure in atm, so use $R = 0.0821$ L·atm/mol·K.

$$D = \frac{PM}{RT} = \frac{3.10 \text{ atm} \times 80.91 \text{ g/mol}}{0.0821 \text{ L·atm/mol·K} \times 268 \text{ K}} = 11.4 \text{ g/L}$$

**EVALUATE**

Are the units correct?

**Yes; units canceled to give g/L, the correct units for gas density.**

Is the number of significant figures correct?

**Yes; the number of significant figures is correct because data were given to three significant figures.**

Is the answer reasonable?

**Yes; 11.4 g/L is a reasonable density for a heavy gas compressed to 3 atm. The calculation can be approximated as 3 × 80/(0.08 × 270) = 3 × 1000/270 = 11.**

# Practice

1. What is the density of silicon tetrafluoride gas at 72°C and a pressure of 144.5 kPa? **ans: 5.24 g/L**

2. At what temperature will nitrogen gas have a density of 1.13 g/L at a pressure of 1.09 atm? **ans: 329 K or 56°C**

| Problem Solving *continued*

# Additional Problems

1. Use the ideal-gas-law equation to calculate the unknown quantity in each of the following sets of measurements.

| | $P$ | $V$ | $n$ | $t$ |
|---|---|---|---|---|
| **a.** | 0.0477 atm | 15 200 L | ? mol | $-15°C$ |
| **b.** | ? kPa | 0.119 mL | 0.000 350 mol | 0°C |
| **c.** | 500.0 kPa | 250. mL | 0.120 mol | ?°C |
| **d.** | 19.5 atm | ? | $4.7 \times 10^4$ mol | 300.°C |

2. Use the ideal-gas-law equation to calculate the unknown quantity in each of the following sets of measurements.

| | $P$ | $V$ | $m$ | $M$ | $t$ |
|---|---|---|---|---|---|
| **a.** | 0.955 atm | 3.77 L | 8.23 g | ? g/mol | 25°C |
| **b.** | 105.0 kPa | 50.0 mL | ? g | 48.02 g/mol | 0°C |
| **c.** | 0.782 atm | ? L | $3.20 \times 10^{-3}$ g | 2.02 g/mol | $-5°C$ |
| **d.** | ? atm | 2.00 L | 7.19 g | 159.8 g/mol | 185°C |
| **e.** | 107.2 kPa | 26.1 mL | 0.414 g | ? g/mol | 45°C |

3. Determine the volume of one mole of an ideal gas at 25°C and 0.915 kPa.

4. Calculate the unknown quantity in each of the following sets of measurements.

| | $P$ | *Molar mass* | *Density* | $t$ |
|---|---|---|---|---|
| **a.** | 1.12 atm | ? g/mol | 2.40 g/L | 2°C |
| **b.** | 7.50 atm | 30.07 g/mol | ? g/L | 20.°C |
| **c.** | 97.4 kPa | 104.09 g/mol | 4.37 g/L | ?°C |
| **d.** | ? atm | 77.95 g/mol | 6.27 g/L | 66°C |

5. What pressure in atmospheres will 1.36 kg of $N_2O$ gas exert when it is compressed in a 25.0 L cylinder and is stored in an outdoor shed where the temperature can reach 59°C during the summer?

6. Aluminum chloride sublimes at high temperatures. What density will the vapor have at 225°C and 0.939 atm pressure?

7. An unknown gas has a density of 0.0262 g/mL at a pressure of 0.918 atm and a temperature of 10.°C. What is the molar mass of the gas?

8. A large balloon contains 11.7 g of helium. What volume will the helium occupy at an altitude of 10 000 m, where the atmospheric pressure is 0.262 atm and the temperature is $-50.°C$?

9. A student collects ethane by water displacement at a temperature of 15°C (vapor pressure of water is 1.5988 kPa) and a total pressure of 100.0 kPa. The volume of the collection bottle is 245 mL. How many moles of ethane are in the bottle?

10. A reaction yields 3.75 L of nitrogen monoxide. The volume is measured at 19°C and at a pressure of 1.10 atm. What mass of NO was produced by the reaction?

11. A reaction has a theoretical yield of 8.83 g of ammonia. The reaction gives off 10.24 L of ammonia measured at 52°C and 105.3 kPa. What was the percent yield of the reaction?

12. An unknown gas has a density of 0.405 g/L at a pressure of 0.889 atm and a temperature of 7°C. Calculate its molar mass.

13. A paper label has been lost from an old tank of compressed gas. To help identify the unknown gas, you must calculate its molar mass. It is known that the tank has a capacity of 90.0 L and weighs 39.2 kg when empty. You find its current mass to be 50.5 kg. The gauge shows a pressure of 1780 kPa when the temperature is 18°C. What is the molar mass of the gas in the cylinder?

14. What is the pressure inside a tank that has a volume of $1.20 \times 10^3$ L and contains 12.0 kg of HCl gas at a temperature of 18°C?

15. What pressure in kPa is exerted at a temperature of 20.°C by compressed neon gas that has a density of 2.70 g/L?

16. A tank with a volume of 658 mL contains 1.50 g of neon gas. The maximum safe pressure that the tank can withstand is $4.50 \times 10^2$ kPa. At what temperature will the tank have that pressure?

17. The atmospheric pressure on Mars is about 6.75 millibars (1 bar = 100 kPa = 0.9869 atm), and the nighttime temperature can be about −75°C on the same day that the daytime temperature goes up to −8°C. What volume would a bag containing 1.00 g of $H_2$ gas have at both the daytime and nighttime temperatures?

18. What is the pressure in kPa of 3.95 mol of $Cl_2$ gas if it is compressed in a cylinder with a volume of 850. mL at a temperature of 15°C?

19. What volume in mL will 0.00660 mol of hydrogen gas occupy at a pressure of 0.907 atm and a temperature of 9°C?

20. What volume will 8.47 kg of sulfur dioxide gas occupy at a pressure of 89.4 kPa and a temperature of 40.°C?

21. A cylinder contains 908 g of compressed helium. It is to be used to inflate a balloon to a final pressure of 128.3 kPa at a temperature of 2°C. What will the volume of the balloon be under these conditions?

22. The density of dry air at 27°C and 100.0 kPa is 1.162 g/L. Use this information to calculate the molar mass of air (calculate as if air were a pure substance).

# Problem Solving

## Stoichiometry of Gases

Now that you have worked with relationships among moles, mass, and volumes of gases, you can easily put these to work in stoichiometry calculations. Many reactions have gaseous reactants, gaseous products, or both.

Reactants and products that are not gases are usually measured in grams or kilograms. As you know, you must convert these masses to amounts in moles before you can relate the quantities by using a balanced chemical equation. Gaseous products and reactants can be related to solid or liquid products and reactants by using the mole ratio, just as solids and liquids are related to each other.

Reactants and products that are gases are usually measured in liters. If the gas is measured at STP, you will need only Avogadro's law to relate the volume and amount of a gas. One mole of any gas at STP occupies 22.4 L. If the gas is not at STP, you will need to use the ideal gas law to determine the number of moles. Once volume has been converted to amount in moles you can use the mole ratios of products and reactants to solve stoichiometry problems involving multiple phases of products and reactants.

$$n = \frac{PV}{RT}$$

If the problem which you are trying to solve involves only gases, there is a simpler way of dealing with the stoichiometric amounts. Look again at the expression for the ideal gas law above; the molar amount of a gas is directly related to its volume. Therefore, the mole ratios of gases given by the coefficients in the balanced equation can be used as volume ratios of those gases to solve stoichiometry problems. No conversion from volume to amount is required to determine the volume of one gas from the volume of another gas in a balanced chemical equation.

There is one condition that must be observed. Gas volumes can be related by mole ratios only when the volumes are measured under the same conditions of temperature and pressure. If they are not, then the volume of one of the gases must be converted to the conditions of the other gas. Usually you will need to use the combined gas law for this conversion.

$$V_2 = \frac{V_1 P_1 T_2}{T_1 P_2}$$

## Problem Solving continued

### General Plan for Solving Gas Stoichiometry Problems

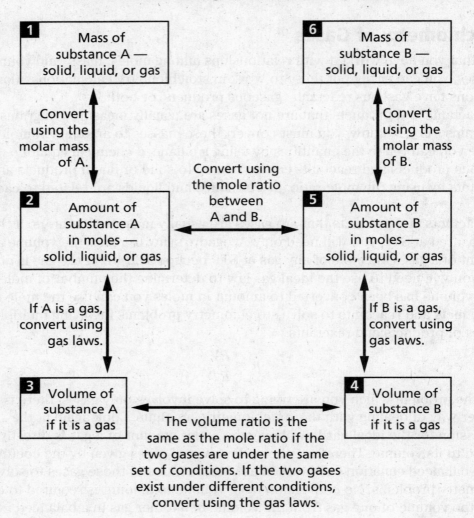

1 | Mass of substance A — solid, liquid, or gas

Convert using the molar mass of A.

2 | Amount of substance A in moles — solid, liquid, or gas

Convert using the mole ratio between A and B.

If A is a gas, convert using gas laws.

3 | Volume of substance A if it is a gas

The volume ratio is the same as the mole ratio if the two gases are under the same set of conditions. If the two gases exist under different conditions, convert using the gas laws.

6 | Mass of substance B — solid, liquid, or gas

Convert using the molar mass of B.

5 | Amount of substance B in moles — solid, liquid, or gas

If B is a gas, convert using gas laws.

4 | Volume of substance B if it is a gas

# Sample Problem 1

**Ammonia can react with oxygen to produce nitrogen and water according to the following equation.**

$$4NH_3(g) + 3O_2(g) \rightarrow 2N_2(g) + 6H_2O(l)$$

**If 1.78 L of $O_2$ reacts, what volume of nitrogen will be produced? Assume that temperature and pressure remain constant.**

# Solution
## ANALYZE

What is given in the problem?    **the balanced equation, the volume of oxygen, and the fact that the two gases exist under the same conditions**

What are you asked to find?    **the volume of $N_2$ produced**

| Items | Data | |
|---|---|---|
| Substance | $O_2$ | $N_2$ |
| Coefficient in balanced equation | 3 | 2 |
| Molar mass | NA | NA |
| Moles | NA | NA |
| Mass of substance | NA | NA |
| Volume of substance | 1.78 L | ? L |
| Temperature conditions | NA | NA |
| Pressure conditions | NA | NA |

## PLAN

What steps are needed to calculate the volume of $N_2$ formed from a given volume of $O_2$?

The coefficients of the balanced equation indicate the mole ratio of $O_2$ to $N_2$. The volume ratio is the same as the mole ratio when volumes are measured under the same conditions.

**3**
Volume of $O_2$
in L

multiply by the volume ratio,
$\frac{N_2}{O_2}$

**4**
Volume of $N_2$
in L

volume ratio, $\frac{N_2}{O_2}$

$\overset{given}{L\ O_2} \times \dfrac{2\ L\ N_2}{3\ L\ O_2} = L\ N_2$

## COMPUTE

$$1.78 \cancel{L\,O_2} \times \frac{2\text{ L N}_2}{3 \cancel{L\,O_2}} = 1.19\text{ L N}_2$$

## EVALUATE

Are the units correct?
**Yes; units canceled to give L $N_2$.**

Is the number of significant figures correct?
**Yes; the number of significant figures is correct because the data were given to three significant figures.**

Is the answer reasonable?
**Yes; the volume of $N_2$ should be 2/3 the volume of $O_2$.**

# Practice

1. In one method of manufacturing nitric acid, ammonia is oxidized to nitrogen monoxide and water.

$$4NH_3(g) + 5O_2(g) \rightarrow 4NO(g) + 6H_2O(l)$$

What volume of oxygen will be used in a reaction of 2800 L of $NH_3$? What volume of NO will be produced? All volumes are measured under the same conditions. **ans: 3500 L $O_2$, 2800 L NO**

2. Fluorine gas reacts violently with water to produce hydrogen fluoride and ozone according to the following equation.

$$3F_2(g) + 3H_2O(l) \rightarrow 6HF(g) + O_3(g)$$

What volumes of $O_3$ and HF gas would be produced by the complete reaction of $3.60 \times 10^4$ mL of fluorine gas? All gases are measured under the same conditions. **ans: $1.20 \times 10^4$ mL $O_3$, $7.20 \times 10^4$ mL HF**

**Problem Solving** *continued*

## Sample Problem 2

Ethylene gas burns in air according to the following equation.

$$C_2H_4(g) + 3O_2(g) \rightarrow 2CO_2(g) + 2H_2O(l)$$

If 13.8 L of $C_2H_4$ measured at 21°C and 1.038 atm burns completely with oxygen, calculate the volume of $CO_2$ produced, assuming the $CO_2$ is measured at 44°C and 0.989 atm.

## Solution

### ANALYZE

What is given in the problem?    the balanced equation, the volume of ethylene, the conditions under which the ethylene was measured, and the conditions under which the $CO_2$ is measured

What are you asked to find?    the volume of $CO_2$ produced as measured at the specified conditions

| Items | Data | |
|---|---|---|
| Substance | $C_2H_4$ | $CO_2$ |
| Coefficient in balanced equation | 1 | 2 |
| Molar mass | NA | NA |
| Moles | NA | NA |
| Mass of substance | NA | NA |
| Volume of substance | 13.8 L | ? L |
| Temperature conditions | 21°C = 294 K | 44°C = 317 K |
| Pressure conditions | 1.083 atm | 0.989 atm |

### PLAN

What steps are needed to calculate the volume of $CO_2$ formed from the complete burning of a given volume of $C_2H_4$?

Use the volume ratio of $C_2H_4$ to $CO_2$ to calculate the volume of $CO_2$ at the same conditions as $C_2H_4$. Convert to the volume of $CO_2$ for the given conditions using the combined gas law.

**▎Problem Solving** *continued*

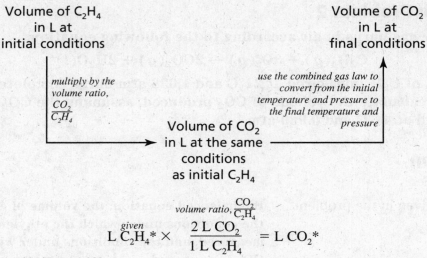

Volume of $C_2H_4$
in L at
initial conditions

Volume of $CO_2$
in L at
final conditions

*multiply by the volume ratio,* $\frac{CO_2}{C_2H_4}$

*use the combined gas law to convert from the initial temperature and pressure to the final temperature and pressure*

Volume of $CO_2$
in L at the same
conditions
as initial $C_2H_4$

*volume ratio,* $\frac{CO_2}{C_2H_4}$

$$\text{L } \overset{given}{C_2H_4}* \times \quad \frac{2 \text{ L CO}_2}{1 \text{ L C}_2H_4} = \text{L CO}_2*$$

\* at 294 K and 1.083 atm

Neither pressure nor temperature is constant; therefore, the combined gas law must be used to calculate the volume of $CO_2$ at the final temperature and pressure.

$$\frac{P_1 V_1}{T_1} = \frac{P_2 V_2}{T_2}$$

$$\frac{\overset{given}{T_2} \times \overset{given}{P_1} \times \overset{\substack{calculated \\ above}}{V_1}}{\underset{given}{P_2} \times \underset{given}{T_1}} = V_2$$

## COMPUTE

$$13.8 \text{ L } C_2H_4* \times \frac{2 \text{ L CO}_2}{1 \text{ L C}_2H_4} = 27.6 \text{ L CO}_2*$$

\* at 294 K and 1.083 atm

Solve the combined-gas-law equation for $V_2$.

$$V_2 = \frac{317 \text{ K} \times 1.083 \text{ atm} \times 27.6 \text{ L CO}_2}{0.989 \text{ atm} \times 294 \text{ K}} = 32.6 \text{ L CO}_2$$

## EVALUATE

Are the units correct?
**Yes; units canceled to give L $CO_2$.**

Is the number of significant figures correct?
**Yes; the number of significant figures is correct because the data had a minimum of three significant figures.**

Is the answer reasonable?
**Yes; the changes in both pressure and temperature increased the volume by small factors.**

**Problem Solving** *continued*

# Practice

**1.** A sample of ethanol burns in $O_2$ to form $CO_2$ and $H_2O$ according to the following equation.

$$C_2H_5OH + 3O_2 \rightarrow 2CO_2 + 3H_2O$$

If the combustion uses 55.8 mL of oxygen measured at 2.26 atm and 40.°C, what volume of $CO_2$ is produced when measured at STP? **ans: 73.3 mL CO$_2$**

**2.** Dinitrogen pentoxide decomposes into nitrogen dioxide and oxygen. If 5.00 L of $N_2O_5$ reacts at STP, what volume of $NO_2$ is produced when measured at 64.5°C and 1.76 atm? **ans: 7.02 L**

**| Problem Solving** *continued*

## Sample Problem 3

**When arsenic(III) sulfide is roasted in air, it reacts with oxygen to produce arsenic(III) oxide and sulfur dioxide according to the following equation.**

$$2As_2S_3(s) + 9O_2(g) \rightarrow 2As_2O_3(s) + 6SO_2(g)$$

**When 89.5 g of $As_2S_3$ is roasted with excess oxygen, what volume of $SO_2$ is produced? The gaseous product is measured at 20°C and 98.0 kPa.**

## Solution
### ANALYZE

What is given in the problem?    **the balanced equation, the mass of $As_2S_3$, and the pressure and temperature conditions under which the $SO_2$ is measured**

What are you asked to find?    **the volume of $SO_2$ produced as measured at the given conditions**

| Items | Data | |
|---|---|---|
| Substance | $As_2S_3(s)$ | $SO_2(g)$ |
| Coefficient in balanced equation | 2 | 6 |
| Molar mass* | 246.05 g/mol | NA |
| Mass of substance | 89.5 g | NA |
| Amount | ? mol | ? mol |
| Volume of substance | NA | ? L |
| Temperature conditions | NA | 20°C = 293 K |
| Pressure conditions | NA | 98.0 kPa |

*determined from the periodic table

### PLAN

What steps are needed to calculate the volume of $SO_2$ formed from the reaction of a given mass of $As_2S_3$?

**Use the molar mass of $As_2S_3$ to determine the number of moles that react. Use the mole ratio from the balanced chemical equation to determine the amount in moles of $SO_2$ formed. Use the ideal-gas-law equation to determine the volume of $SO_2$ formed from the amount in moles.**

| Problem Solving *continued*

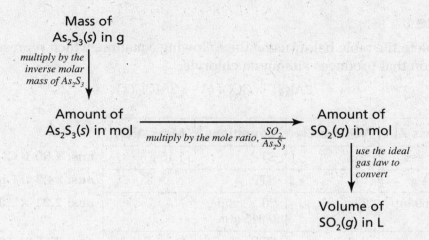

$$\underset{given}{\text{g As}_2\text{S}_3} \times \underset{\substack{\frac{1}{\text{molar mass As}_2\text{S}_3}}}{\frac{1 \text{ mol As}_2\text{S}_3}{246.05 \text{ g As}_2\text{S}_3}} \times \underset{\substack{\text{mole ratio, } \frac{\text{SO}_2}{\text{As}_2\text{S}_3}}}{\frac{6 \text{ mol SO}_2}{2 \text{ mol As}_2\text{S}_3}} = \text{mol SO}_2$$

Rearrange the ideal-gas-law equation to solve for the unknown quantity, $V$.

$$PV = nRT$$

$$V = \frac{nRT}{P}$$

$$\frac{\underset{\substack{calculated\ above}}{\text{mol SO}_2} \times \underset{\substack{ideal\ gas\ constant}}{8.314 \text{ L·kPa/mol·K}} \times \underset{\substack{given}}{\text{K}}}{\underset{given}{\text{kPa}}} = \text{L SO}_2$$

## COMPUTE

$$89.5 \text{ g As}_2\text{S}_3 \times \frac{1 \text{ mol As}_2\text{S}_3}{246.05 \text{ g As}_2\text{S}_3} \times \frac{6 \text{ mol SO}_2}{2 \text{ mol As}_2\text{S}_3} = 1.09 \text{ mol SO}_2$$

$$\frac{1.09 \text{ mol SO}_2 \times 8.314 \text{ L·kPa/mol·K} \times 293 \text{ K}}{98.0 \text{ kPa}} = 27.1 \text{ L SO}_2$$

## EVALUATE

Are the units correct?

**Yes; units canceled to give liters of $SO_2$.**

Is the number of significant figures correct?

**Yes; the number of significant figures is correct because the data had a minimum of three significant figures.**

Is the answer reasonable?

**Yes; computation of the amount of $SO_2$ can be approximated as $(9/25) \times 3 = 27/25$, so you would expect an answer a little greater than 1. At a temperature slightly above standard temperature, you would expect a volume a little greater than 22.4 L.**

**| Problem Solving** *continued*

## Practice

**1.** Complete the table below using the following equation, which represents a reaction that produces aluminum chloride.

$$2Al(s) + 3Cl_2(g) \rightarrow 2AlCl_3(s)$$

|     | Mass Al   | Volume Cl$_2$ | Conditions            | Mass AlCl$_3$ |                                       |
| --- | --------- | ------------- | --------------------- | ------------- | ------------------------------------- |
| **a.** | excess    | ? L           | STP                   | 7.15 g        | ans: 1.80 L Cl$_2$                    |
| **b.** | 19.4 g    | ? L           | STP                   | NA            | ans: 24.2 L Cl$_2$                    |
| **c.** | 1.559 kg  | ? L           | 20.°C and 0.945 atm   | NA            | ans: 2.21 × 10$^3$ L Cl$_2$           |
| **d.** | excess    | 920. L        | STP                   | ? g           | ans: 3.65 × 10$^3$ g AlCl$_3$         |
| **e.** | ? g       | 1.049 mL      | 37°C and 5.00 atm     | NA            | ans: 3.71 × 10$^{-3}$ g Al            |
| **f.** | 500.00 kg | ? m$^3$       | 15°C and 83.0 kPa     | NA            | ans: 8.02 × 10$^2$ m$^3$ Cl$_2$       |

# Additional Problems

**1.** The industrial production of ammonia proceeds according to the following equation.

$$N_2(g) + 3H_2(g) \rightarrow 2NH_3(g)$$

**a.** What volume of nitrogen at STP is needed to react with 57.0 mL of hydrogen measured at STP?

**b.** What volume of $NH_3$ at STP can be produced from the complete reaction of $6.39 \times 10^4$ L of hydrogen?

**c.** If 20.0 mol of nitrogen is available, what volume of $NH_3$ at STP can be produced?

**d.** What volume of $H_2$ at STP will be needed to produce 800. L of ammonia, measured at 55°C and 0.900 atm?

**2.** Propane burns according to the following equation.

$$C_3H_8(g) + 5O_2(g) \rightarrow 3CO_2(g) + 4H_2O(g)$$

**a.** What volume of water vapor measured at 250.°C and 1.00 atm is produced when 3.0 L of propane at STP is burned?

**b.** What volume of oxygen at 20.°C and 102.6 kPa is used if 640. L of $CO_2$ is produced? The $CO_2$ is also measured at 20.°C and 102.6 kPa.

**c.** If 465 mL of oxygen at STP is used in the reaction, what volume of $CO_2$, measured at 37°C and 0.973 atm, is produced?

**d.** When 2.50 L of $C_3H_8$ at STP burns, what total volume of gaseous products is formed? The volume of the products is measured at 175°C and 1.14 atm.

**3.** Carbon monoxide will burn in air to produce $CO_2$ according to the following equation.

$$2CO(g) + O_2(g) \rightarrow 2CO_2(g)$$

What volume of oxygen at STP will be needed to react with 3500. L of CO measured at 20.°C and a pressure of 0.953 atm?

**4.** Silicon tetrafluoride gas can be produced by the action of HF on silica according to the following equation.

$$SiO_2(s) + 4HF(g) \rightarrow SiF_4(g) + 2H_2O(l)$$

1.00 L of HF gas under pressure at 3.48 atm and a temperature of 25°C reacts completely with $SiO_2$ to form $SiF_4$. What volume of $SiF_4$, measured at 15°C and 0.940 atm, is produced by this reaction?

**5.** One method used in the eighteenth century to generate hydrogen was to pass steam through red-hot steel tubes. The following reaction takes place.

$$3Fe(s) + 4H_2O(g) \rightarrow Fe_3O_4(s) + 4H_2(g)$$

**a.** What volume of hydrogen at STP can be produced by the reaction of 6.28 g of iron?

**b.** What mass of iron will react with 500. L of steam at 250.°C and 1.00 atm pressure?

**c.** If 285 g of $Fe_3O_4$ are formed, what volume of hydrogen, measured at 20.°C and 1.06 atm, is produced?

**6.** Sodium reacts vigorously with water to produce hydrogen and sodium hydroxide according to the following equation.

$$2Na(s) + 2H_2O(l) \rightarrow 2NaOH(aq) + H_2(g)$$

If 0.027 g of sodium reacts with excess water, what volume of hydrogen at STP is formed?

**7.** Diethyl ether burns in air according to the following equation.

$$C_4H_{10}O(l) + 6O_2(g) \rightarrow 4CO_2(g) + 5H_2O(l)$$

If 7.15 L of $CO_2$ is produced at a temperature of 125°C and a pressure of 1.02 atm, what volume of oxygen, measured at STP, was consumed and what mass of diethyl ether was burned?

**8.** When nitroglycerin detonates, it produces large volumes of hot gases almost instantly according to the following equation.

$$4C_3H_5N_3O_9(l) \rightarrow 6N_2(g) + 12CO_2(g) + 10H_2O(g) + O_2(g)$$

**a.** When 0.100 mol of nitroglycerin explodes, what volume of each gas measured at STP is produced?

**b.** What total volume of gases is produced at 300.°C and 1.00 atm when 10.0 g of nitroglycerin explodes?

**9.** Dinitrogen monoxide can be prepared by heating ammonium nitrate, which decomposes according to the following equation.

$$NH_4NO_3(s) \rightarrow N_2O(g) + 2H_2O(l)$$

What mass of ammonium nitrate should be decomposed in order to produce 250. mL of $N_2O$, measured at STP?

**10.** Phosphine, $PH_3$, is the phosphorus analogue to ammonia, $NH_3$. It can be produced by the reaction between calcium phosphide and water according to the following equation.

$$Ca_3P_2(s) + 6H_2O(l) \rightarrow 3Ca(OH)_2(s \text{ and } aq) + 2PH_3(g)$$

What volume of phosphine, measured at 18°C and 102.4 kPa, is produced by the reaction of 8.46 g of $Ca_3P_2$?

**11.** In one method of producing aluminum chloride, HCl gas is passed over aluminum and the following reaction takes place.

$$2Al(s) + 6HCl(g) \rightarrow 2AlCl_3(g) + 3H_2(g)$$

What mass of Al should be on hand in order to produce $6.0 \times 10^3$ kg of $AlCl_3$? What volume of compressed HCl at 4.71 atm and a temperature of 43°C should be on hand at the same time?

**▌Problem Solving** *continued*

**12.** Urea, $(NH_2)_2CO$, is an important fertilizer that is manufactured by the following reaction.

$$2NH_3(g) + CO_2(g) \rightarrow (NH_2)_2CO(s) + H_2O(g)$$

What volume of $NH_3$ at STP will be needed to produce $8.50 \times 10^4$ kg of urea if there is an 89.5% yield in the process?

**13.** An obsolete method of generating oxygen in the laboratory involves the decomposition of barium peroxide by the following equation.

$$2BaO_2(s) \rightarrow 2BaO(s) + O_2(g)$$

What mass of $BaO_2$ reacted if 265 mL of $O_2$ is collected by water displacement at 0.975 atm and 10.°C?

**14.** It is possible to generate chlorine gas by dripping concentrated HCl solution onto solid potassium permanganate according to the following equation.

$$2KMnO_4(s) + 16HCl(aq) \rightarrow 2KCl(aq) + 2MnCl_2(aq) + 8H_2O(l) + 5Cl_2(g)$$

If excess HCl is dripped onto 15.0 g of $KMnO_4$, what volume of $Cl_2$ will be produced? The $Cl_2$ is measured at 15°C and 0.959 atm.

**15.** Ammonia can be oxidized in the presence of a platinum catalyst according to the following equation.

$$4NH_3(g) + 5O_2(g) \rightarrow 4NO(g) + 6H_2O(l)$$

The NO that is produced reacts almost immediately with additional oxygen according to the following equation.

$$2NO(g) + O_2(g) \rightarrow 2NO_2(g)$$

If 35.0 kL of oxygen at STP react in the first reaction, what volume of $NH_3$ at STP reacts with it? What volume of $NO_2$ at STP will be formed in the second reaction, assuming there is excess oxygen that was not used up in the first reaction?

**16.** Oxygen can be generated in the laboratory by heating potassium chlorate. The reaction is represented by the following equation.

$$2KClO_3(s) \rightarrow 2KCl(s) + 3O_2(g)$$

What mass of $KClO_3$ must be used in order to generate 5.00 L of $O_2$, measured at STP?

**17.** One of the reactions in the Solvay process is used to make sodium hydrogen carbonate. It occurs when carbon dioxide and ammonia are passed through concentrated salt brine. The following equation represents the reaction.

$$NaCl(aq) + H_2O(l) + CO_2(g) + NH_3(g) \rightarrow NaHCO_3(s) + NH_4Cl(aq)$$

**a.** What volume of $NH_3$ at 25°C and 1.00 atm pressure will be required if 38 000 L of $CO_2$, measured under the same conditions, react to form $NaHCO_3$?

**b.** What mass of $NaHCO_3$ can be formed when the gases in (a) react with NaCl?

**c.** If this reaction forms 46.0 kg of $NaHCO_3$, what volume of $NH_3$, measured at STP, reacted?

**d.** What volume of $CO_2$, compressed in a tank at 5.50 atm and a temperature of 42°C, will be needed to produce 100.00 kg of $NaHCO_3$?

**18.** The combustion of butane is represented in the following equation.

$$2C_4H_{10}(g) + 13O_2(g) \rightarrow 8CO_2(g) + 10H_2O(l)$$

**a.** If 4.74 g of butane react with excess oxygen, what volume of $CO_2$, measured at 150.°C and 1.14 atm, will be formed?

**b.** What volume of oxygen, measured at 0.980 atm and 75°C, will be consumed by the complete combustion of 0.500 g of butane?

**c.** A butane-fueled torch has a mass of 876.2 g. After burning for some time, the torch has a mass of 859.3 g. What volume of $CO_2$, at STP, was formed while the torch burned?

**d.** What mass of $H_2O$ is produced when butane burns and produces 3720 L of $CO_2$, measured at 35°C and 0.993 atm pressure?

Skills Worksheet

# Problem Solving

## Concentration of Solutions

There are three principal ways to express solution concentration in chemistry—
percentage by mass, molarity, and molality.

The following table compares these three ways of stating solution concentration. Examining the method of preparation of the three types may help you understand the differences among them.

|  | Symbol | Meaning | How to prepare |
|---|---|---|---|
| **Percentage** | % | Grams solute per 100 g of solution | **5%:** Dissolve 5 g of solute in 95 g solvent. |
| **Molarity** | M | Moles solute per liter of solution | **5 M:** Dissolve 5 mol of solute in solvent and add solvent to make 1 L of solution. |
| **Molality** | *m* | Moles solute per kilogram of solvent | **5 *m*:** Dissolve 5 mol of solute in 1 kg of solvent. |

## PERCENTAGE CONCENTRATION

You will find percentages of solutes stated on the labels of many commercial products, such as household cleaners, liquid pesticide solutions, and shampoos. If your sink becomes clogged, you might buy a bottle of drain opener whose label states that it is a 2.4% sodium hydroxide solution. This means that the bottle contains 2.4 g of NaOH for every 100 g of solution.

Computing percentage concentration is very much like computing percentage composition. Both involve finding the percentage of a single component of a multicomponent system. In each type of percentage calculation, the mass of the important component (in percentage concentration, the solute) is divided by the total mass of the system and multiplied by 100 to yield a percentage. In percentage concentration, the solute is the important component, and the total mass of the system is the mass of the solute plus the mass of the solvent.

**General Plan for Solving Percentage Concentration Problems**

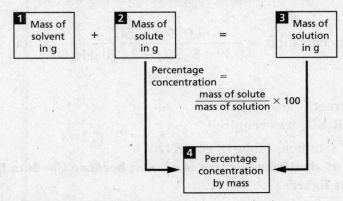

## Sample Problem 1

**What is the percentage by mass of a solution made by dissolving 0.49 g of potassium sulfate in 12.70 g of water?**

## Solution

### ANALYZE

What is given in the problem?   **the mass of solvent, and the mass of solute, $K_2SO_4$**

What are you asked to find?   **the concentration of the solution expressed as a percentage by mass**

| Items | Data |
|-------|------|
| Mass of solvent | 12.70 g $H_2O$ |
| Mass of solute | 0.49 g $K_2SO_4$ |
| Concentration (% by mass) | ? % |

### PLAN

What step is needed to calculate the concentration of the solution as a percentage by mass?

**Divide the mass of solute by the mass of the solution and multiply by 100.**

**1**  **2**  **3**
Mass of water in g + Mass of $K_2SO_4$ in g = Mass of $K_2SO_4$ solution in g

$$percentage\ concentration = \frac{solute\ mass}{solution\ mass} \times 100$$

**4**
Percentage
$K_2SO_4$ by mass

$$percentage\ concentration = \frac{g\ K_2SO_4\ (given)}{g\ K_2SO_4\ (given) + g\ H_2O\ (given)} \times 100$$

### COMPUTE

$$percentage\ concentration = \frac{0.49\ g\ K_2SO_4}{0.49\ g\ K_2SO_4 + 12.70\ g\ H_2O} \times 100 = 3.7\%\ K_2SO_4$$

### EVALUATE

Are the units correct?
**Yes; percentage $K_2SO_4$ was required.**

Is the number of significant figures correct?
**Yes; the number of significant figures is correct because the data had a minimum of two significant figures.**

**| Problem Solving** *continued*

Is the answer reasonable?
**Yes; the computation can be approximated as 0.5/13 × 100 = 3.8%.**

## Practice

**1.** What is the percentage concentration of 75.0 g of ethanol dissolved in 500.0 g of water? **ans: 13.0% ethanol**

**2.** A chemist dissolves 3.50 g of potassium iodate and 6.23 g of potassium hydroxide in 805.05 g of water. What is the percentage concentration of each solute in the solution? **ans: 0.430% KIO$_3$, 0.765% KOH**

**3.** A student wants to make a 5.00% solution of rubidium chloride using 0.377 g of the substance. What mass of water will be needed to make the solution? **ans: 7.16 g H$_2$O**

**4.** What mass of lithium nitrate would have to be dissolved in 30.0 g of water in order to make an 18.0% solution? **ans: 6.59 g LiNO$_3$**

| **Problem Solving** *continued* |

## MOLARITY

Molarity is the most common way to express concentration in chemistry. Molarity is the number of moles of solute per liter of solution and is given as a number followed by a capital M. A 2 M solution of nitric acid contains 2 mol of $HNO_3$ per liter of solution. As you know, substances react in mole ratios. Knowing the molar concentration of a solution allows you to measure a number of moles of a dissolved substance by measuring the volume of solution.

### General Plan for Solving Molarity Problems

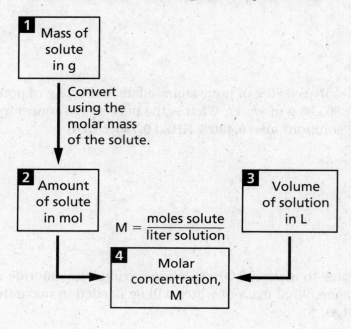

**| Problem Solving** *continued*

# Sample Problem 2

**What is the molarity of a solution prepared by dissolving 37.94 g of potassium hydroxide in some water and then diluting the solution to a volume of 500.00 mL?**

# Solution

## ANALYZE

What is given in the problem?  **the mass of the solute, KOH, and the final volume of the solution**

What are you asked to find?  **the concentration of the solution expressed as molarity**

| Items | Data |
|---|---|
| Mass of solute | 37.94 g KOH |
| Moles of solute | ? mol KOH |
| Molar mass of solute* | 56.11 g/mol |
| Volume of solution | 500.00 mL |
| Concentration (molarity) | ? M |

*determined from the periodic table

## PLAN

What steps are needed to calculate the concentration of the solution as molarity?
**Determine the amount in moles of solute; calculate the moles per liter of solution.**

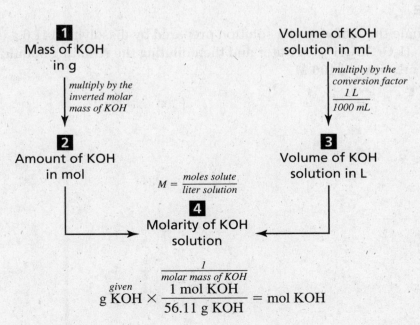

$$g\ \overset{given}{KOH} \times \frac{1\ mol\ KOH}{56.11\ g\ KOH} = mol\ KOH$$

$$\overset{given}{\text{mL solution}} \times \frac{1 \text{ L}}{1000 \text{ mL}} = \text{L solution}$$

$$\frac{\overset{calculated\ above}{\text{mol KOH}}}{\underset{calculated\ above}{\text{L solution}}} = \text{M solution}$$

**COMPUTE**

$$37.94 \text{ g } \cancel{\text{KOH}} \times \frac{1 \text{ mol KOH}}{56.11 \text{ g } \cancel{\text{KOH}}} = 0.6762 \text{ mol KOH}$$

$$500.00 \text{ m} \cancel{\text{L}} \text{ solution} \times \frac{1 \text{ L}}{1000 \text{ m} \cancel{\text{L}}} = 0.500\ 00 \text{ L solution}$$

$$\frac{0.6762 \text{ mol KOH}}{0.500\ 00 \text{ L solution}} = 1.352 \text{ M}$$

**EVALUATE**

Are the units correct?
**Yes; units canceled to give moles KOH per liter of solution.**

Is the number of significant figures correct?
**Yes; the number of significant figures is correct because the data had a minimum of four significant figures.**

Is the answer reasonable?
**Yes; note that 0.6762 mol is approximately 2/3 mol and 0.500 00 L is 1/2 L. Thus, the calculation can be estimated as (2/3)/(1/2) = 4/3, which is very close to the result.**

# Practice

1. Determine the molarity of a solution prepared by dissolving 141.6 g of citric acid, $C_3H_5O(COOH)_3$, in water and then diluting the resulting solution to 3500.0 mL. **ans: 0.2106 M**

**Problem Solving** *continued*

2. What is the molarity of a salt solution made by dissolving 280.0 mg of NaCl in 2.00 mL of water? Assume the final volume is the same as the volume of the water. **ans: 2.40 M**

3. What is the molarity of a solution that contains 390.0 g of acetic acid, $CH_3COOH$, dissolved in enough acetone to make 1000.0 mL of solution? **ans: 6.494 M**

## Sample Problem 3

An analytical chemist wants to make 750.0 mL of a 6.00 M solution of sodium hydroxide. What mass of NaOH will the chemist need to make this solution?

## Solution

### ANALYZE

What is given in the problem?    the identity of the solute, the total volume of solution, and the molarity of the solution

What are you asked to find?    the mass of solute to dissolve

| Items | Data |
|---|---|
| Mass of solute | ? g NaOH |
| Molar mass of solute | 40.00 g/mol |
| Moles of solute | ? mol NaOH |
| Volume of solution | 750.0 mL |
| Concentration (molarity) | 6.00 M |

### PLAN

What steps are needed to calculate the mass of solute needed?

**Determine the amount in moles needed for the solution required, and convert to grams by multiplying by the molar mass of the solute.**

### COMPUTE

$$\text{Molarity of NaOH solution} \times \text{Volume of NaOH solution in L} = \text{Amount of NaOH in mol}$$

multiply by the conversion factor $\dfrac{1\ L}{1000\ mL}$

multiply by the molar mass of NaOH

Volume of NaOH solution in mL

Mass of NaOH in g

$$\overset{given}{\text{mL solution}} \times \frac{1\ L}{1000\ mL} = \text{L solution}$$

$$\overset{given}{\frac{\text{mol NaOH}}{\text{L solution}}} \times \overset{\substack{calculated \\ above}}{\text{L solution}} \times \overset{\substack{molar\ mass\ of\ NaOH}}{\frac{40.00\ \text{g NaOH}}{1\ \text{mol NaOH}}} = \text{g NaOH}$$

$$750.0\ \text{mL solution} \times \frac{1\ L}{1000\ mL} = 0.7500\ \text{L solution}$$

$$\frac{6.00 \text{ mol NaOH}}{\text{L solution}} \times 0.7500 \text{ L solution} \times \frac{40.00 \text{ g NaOH}}{1 \text{ mol NaOH}} = 180. \text{ g NaOH}$$

## EVALUATE

Are the units correct?

**Yes; units canceled to give grams of NaOH.**

Is the number of significant figures correct?

**Yes; the number of significant figures is correct because the data had a minimum of three significant figures.**

Is the answer reasonable?

**Yes; the calculation can be estimated as (3/4) × (6)(40) = (3/4) × 240 = 180.**

# Practice

**1.** What mass of glucose, $C_6H_{12}O_6$, would be required to prepare $5.000 \times 10^3$ L of a 0.215 M solution? **ans: $1.94 \times 10^5$ g**

**2.** What mass of magnesium bromide would be required to prepare 720. mL of a 0.0939 M aqueous solution? **ans: 12.4 g**

**3.** What mass of ammonium chloride is dissolved in 300. mL of a 0.875 M solution? **ans: 14.0 g**

## MOLALITY

Molality is the amount in moles of solute per kilogram of solvent and is given by a number followed by an italic lowercase $m$. A 5 $m$ aqueous solution of glucose contains 5 mol of $C_6H_{12}O_6$ per kilogram of water. Molal concentration is important primarily in working with colligative properties of solutions.

**General Plan for Solving Molality Problems**

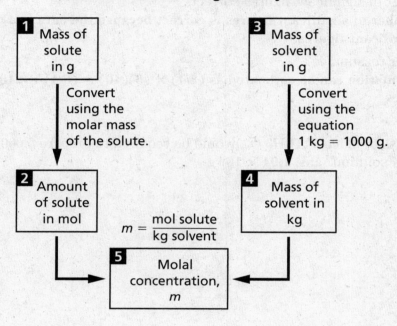

# Sample Problem 4

Determine the molal concentration of a solution containing 81.3 g of
ethylene glycol, $HOCH_2CH_2OH$, dissolved in 166 g of water.

# Solution

## ANALYZE

What is given in the problem?     **the mass of ethylene glycol dissolved, and the
                                   mass of the solvent, water**

What are you asked to find?       **the molal concentration of the solution**

| Items | Data |
|---|---|
| Mass of solute | 81.3 g ethylene glycol |
| Molar mass of solute | 62.08 g/mol ethylene glycol |
| Moles of solute | ? mol ethylene glycol |
| Mass of solvent | 166 g $H_2O$ |
| Concentration (molality) | ? m |

## PLAN

What steps are needed to calculate the molal concentration of the ethylene glycol
solution?

**Determine the amount of solute in moles and the mass of solvent in kilograms;
calculate the moles of solute per kilogram of solvent.**

**1** Mass of $C_2H_6O_2$ in g

*multiply by the
inverted molar
mass of $C_2H_6O_2$*

**2** Amount of $C_2H_6O_2$
in mol

$m = \dfrac{moles\ C_2H_6O_2}{kg\ H_2O}$

**3** Mass of $H_2O$ in g

*multiply by the
conversion factor*
$\dfrac{1\ kg}{1000\ g}$

**4** Mass of $H_2O$ in kg

**5** Molality of $C_2H_6O_2$
solution

$$\overset{given}{g\ C_2H_6O_2} \times \overset{\frac{1}{molar\ mass\ of\ C_2H_6O_2}}{\dfrac{1\ mol\ C_2H_6O_2}{62.08\ g\ C_2H_6O_2}} = mol\ C_2H_6O_2$$

$$\overset{given}{g\ H_2O} \times \dfrac{1\ kg}{1000\ g} = kg\ H_2O$$

$$\dfrac{\overset{calculated\ above}{mol\ C_2H_6O_2}}{\underset{calculated\ above}{kg\ H_2O}} = m\ C_2H_6O_2\ solution$$

**COMPUTE**

$$81.3 \text{ g } C_2H_6O_2 \times \frac{1 \text{ mol } C_2H_6O_2}{62.08 \text{ g } C_2H_6O_2} = 1.31 \text{ mol } C_2H_6O_2$$

$$166 \text{ g } H_2O \times \frac{1 \text{ kg}}{1000 \text{ g}} = 0.166 \text{ kg } H_2O$$

$$\frac{1.31 \text{ mol } C_2H_6O_2}{0.166 \text{ kg } H_2O} = 7.89 \text{ } m$$

**EVALUATE**

Are the units correct?

**Yes; units canceled to give moles $C_2H_6O_2$ per kilogram of solvent.**

Is the number of significant figures correct?

**Yes; the number of significant figures is correct because the data had a minimum of three significant figures.**

Is the answer reasonable?

**Yes; because 1.31 mol is approximately 4/3 mol and 0.166 kg is approximately 1/6 kg, the calculation can be estimated as (4/3)/(1/6) = 24/3 = 8, which is very close to the result.**

## Practice

**1.** Determine the molality of a solution of 560 g of acetone, $CH_3COCH_3$, in 620 g of water. **ans: 16 $m$**

**2.** What is the molality of a solution of 12.9 g of fructose, $C_6H_{12}O_6$, in 31.0 g of water? **ans: 2.31 $m$**

**3.** How many moles of 2-butanol, $CH_3CHOHCH_2CH_3$, must be dissolved in 125 g of ethanol in order to produce a 12.0 $m$ 2-butanol solution? What mass of 2-butanol is this? **ans: 1.50 mol 2-butanol, 111 g 2-butanol**

| Problem Solving *continued*

# Additional Problems

1. Complete the table below by determining the missing quantity in each example. All solutions are aqueous. Any quantity that is not applicable to a given solution is marked NA.

|    | Solution made | Mass of solute used | Quantity of solution made | Quantity of solvent used |
|----|---------------|---------------------|---------------------------|--------------------------|
| a. | 12.0% $KMnO_4$ | ? g $KMnO_4$ | 500.0 g | ? g $H_2O$ |
| b. | 0.60 M $BaCl_2$ | ? g $BaCl_2$ | 1.750 L | NA |
| c. | 6.20 $m$ glycerol, $HOCH_2CHOHCH_2OH$ | ? g glycerol | NA | 800.0 g $H_2O$ |
| d. | ? M $K_2Cr_2O_7$ | 12.27 g $K_2Cr_2O_7$ | 650. mL | NA |
| e. | ? $m$ $CaCl_2$ | 288 g $CaCl_2$ | NA | 2.04 kg $H_2O$ |
| f. | 0.160 M NaCl | ? g NaCl | 25.0 mL | NA |
| g. | 2.00 $m$ glucose, $C_6H_{12}O_6$ | ? g glucose | ? g solution | 1.50 kg $H_2O$ |

2. How many moles of $H_2SO_4$ are in 2.50 L of a 4.25 M aqueous solution?

3. Determine the molal concentration of 71.5 g of linoleic acid, $C_{18}H_{32}O_2$, in 525 g of hexane, $C_6H_{14}$.

4. You have a solution that is 16.2% sodium thiosulfate, $Na_2S_2O_3$, by mass.

   a. What mass of sodium thiosulfate is in 80.0 g of solution?

   b. How many moles of sodium thiosulfate are in 80.0 g of solution?

   c. If 80.0 g of the sodium thiosulfate solution is diluted to 250.0 mL with water, what is the molarity of the resulting solution?

5. What mass of anhydrous cobalt(II) chloride would be needed in order to make 650.00 mL of a 4.00 M cobalt(II) chloride solution?

6. A student wants to make a 0.150 M aqueous solution of silver nitrate, $AgNO_3$ and has a bottle containing 11.27 g of silver nitrate. What should be the final volume of the solution?

7. What mass of urea, $NH_2CONH_2$, must be dissolved in 2250 g of water in order to prepare a 1.50 $m$ solution?

8. What mass of barium nitrate is dissolved in 21.29 mL of a 3.38 M solution?

9. Describe what you would do to prepare 100.0 g of a 3.5% solution of ammonium sulfate in water.

10. What mass of anhydrous calcium chloride should be dissolved in 590.0 g of water in order to produce a 0.82 $m$ solution?

11. How many moles of ammonia are in 0.250 L of a 5.00 M aqueous ammonia solution? If this solution were diluted to 1.000 L, what would be the molarity of the resulting solution?

**12.** What is the molar mass of a solute if 62.0 g of the solute in 125 g of water produce a 5.3 $m$ solution?

**13.** A saline solution is 0.9% NaCl. What masses of NaCl and water would be required to prepare 50. L of this saline solution? Assume that the density of water is 1.000 g/mL and that the NaCl does not add to the volume of the solution.

**14.** A student weighs an empty beaker on a balance and finds its mass to be 68.60 g. The student weighs the beaker again after adding water and finds the new mass to be 115.12 g. A mass of 4.08 g of glucose is then dissolved in the water. What is the percentage concentration of glucose in the solution?

**15.** The density of ethyl acetate at 20°C is 0.902 g/mL. What volume of ethyl acetate at 20°C would be required to prepare a 2.0% solution of cellulose nitrate using 25 g of cellulose nitrate?

**16.** Aqueous cadmium chloride reacts with sodium sulfide to produce bright-yellow cadmium sulfide. Write the balanced equation for this reaction and answer the following questions.

**a.** How many moles of $CdCl_2$ are in 50.00 mL of a 3.91 M solution?

**b.** If the solution in (a) reacted with excess sodium sulfide, how many moles of CdS would be formed?

**c.** What mass of CdS would be formed?

**17.** What mass of $H_2SO_4$ is contained in 60.00 mL of a 5.85 M solution of sulfuric acid?

**18.** A truck carrying 22.5 kL of 6.83 M aqueous hydrochloric acid used to clean brick and masonry has overturned. The authorities plan to neutralize the acid with sodium carbonate. How many moles of HCl will have to be neutralized?

**19.** A chemist wants to produce 12.00 g of barium sulfate by reacting a 0.600 M $BaCl_2$ solution with excess $H_2SO_4$, as shown in the reaction below. What volume of the $BaCl_2$ solution should be used?

$$BaCl_2 + H_2SO_4 \rightarrow BaSO_4 + 2HCl$$

**20.** Many substances are hydrates. Whenever you make a solution, it is important to know whether or not the solute you are using is a hydrate and, if it is a hydrate, how many molecules of water are present per formula unit of the substance. This water must be taken into account when weighing out the solute. Something else to remember when making aqueous solutions from hydrates is that once the hydrate is dissolved, the water of hydration is considered to be part of the solvent. A common hydrate used in the chemistry laboratory is copper sulfate pentahydrate, $CuSO_4 \cdot 5H_2O$. Describe how you would make each of the following solutions using $CuSO_4 \cdot 5H_2O$. Specify masses and volumes as needed.

**a.** 100. g of a 6.00% solution of $CuSO_4$

**b.** 1.00 L of a 0.800 M solution of $CuSO_4$

**c.** a 3.5 $m$ solution of $CuSO_4$ in 1.0 kg of water

**| Problem Solving** *continued*

**21.** What mass of calcium chloride hexahydrate is required in order to make 700.0 mL of a 2.50 M solution?

**22.** What mass of the amino acid arginine, $C_6H_{14}N_4O_2$, would be required to make 1.250 L of a 0.00205 M solution?

**23.** How much water would you have to add to 2.402 kg of nickel(II) sulfate hexahydrate in order to prepare a 25.00% solution?

**24.** What mass of potassium aluminum sulfate dodecahydrate, $KAl(SO_4)_2 \cdot 12H_2O$, would be needed to prepare 35.00 g of a 15.00% $KAl(SO_4)_2$ solution? What mass of water would be added to make this solution?

Skills Worksheet

# Problem Solving

## Dilutions

Suppose you work in the laboratory of a paint company where you use 100 mL of a 0.1 M solution of zinc chloride in a quality-control test that you carry out 10 times a day. It would be tedious and time-consuming to continually measure out small amounts of $ZnCl_2$ to make 100 mL of this solution. Of course, you could make many liters of the solution at one time, but that would require several large containers to store the solution. The answer to the problem is to make a much more concentrated solution and then dilute it with water to make the less concentrated solution that you need. The more-concentrated solution is called a *stock solution*. You could make a 1 M $ZnCl_2$ solution by measuring out 1 mol of zinc chloride, 136.3 g, and dissolving it in enough water to make a liter of solution. This solution is 10 times as concentrated as the solution you need. Every time you need the test solution, you can measure out 10 mL of the 1 M solution and dilute it to 100 mL to yield 100 mL of 0.1 M $ZnCl_2$ solution.

To make a solution by dilution, you must determine the volume of stock solution to use and the amount of solvent needed to dilute to the concentration you need. As you have learned, the molarity of a solution is its concentration in moles of solute per liter of solution. Molarity is found by dividing the moles of solute by the number of liters of solution.

$$M = \frac{\text{moles solute}}{\text{liter solution}}$$

So, for a measured volume of any solution:

*amount of solute in mol = molarity × volume of solution*

If this measured volume of solution is diluted to a new volume by adding solvent, the new, larger volume still contains the same number of moles of solute. Therefore, where 1 and 2 represent the concentrated and diluted solutions:

*molarity₁ × volume₁ = moles solute = molarity₂ × volume₂*

Therefore:

*molarity₁ × volume₁ = molarity₂ × volume₂*

This relationship applies whenever solution 2 is made from solution 1 by dilution.

**Problem Solving** *continued*

## General Plan for Solving Dilution Problems

$$\boxed{1 \quad M_1 V_1 = M_2 V_2}$$

Rearrange the equation
$M_1 V_1 = M_2 V_2$
algebraically to solve for
the unknown quantity.

$$\boxed{\begin{array}{c} 2 \\ \text{The equation used to} \\ \text{calculate the unknown quantity} \\ \text{will be one of the following four:} \\[6pt] V_2 = \dfrac{M_1 V_1}{M_2}, \quad M_2 = \dfrac{M_1 V_1}{V_2}, \quad V_1 = \dfrac{M_2 V_2}{M_1}, \quad M_1 = \dfrac{M_2 V_2}{V_1} \end{array}}$$

Substitute each of the
known quantities for its
symbol, and calculate.

$$\boxed{\begin{array}{c} 3 \\ \text{Unknown} \\ \text{molarity or volume} \end{array}}$$

## Sample Problem 1

**What is the molarity of a solution that is made by diluting 50.00 mL of a 4.74 M solution of HCl to 250.00 mL?**

## Solution

### ANALYZE

What is given in the problem?  **the molarity of the stock solution, the volume used to dilute, and the volume of the diluted solution**

What are you asked to find?  **the molarity of the diluted solution**

| Items | Data |
|---|---|
| Concentration of the stock solution ($M_1$) | 4.74 M HCl |
| Volume of stock solution used ($V_1$) | 50.00 mL |
| Volume of diluted solution ($V_2$) | 250.00 mL |
| Concentration of the diluted solution ($M_2$) | ? M |

### PLAN

What step is needed to calculate the concentration of the diluted solution?
**Apply the principle that *volume₁ × molarity₁ = volume₂ × molarity₂.***

**1**
$$M_1V_1 = M_2V_2$$
*rearrange the equation*
$$M_1V_1 = M_2V_2$$
*algebraically to solve for $M_2$*
*substitute each of the known quantities for its symbol and calculate*
**2**
$$M_2 = \frac{M_1V_1}{V_2}$$

*given*
$$M_2 = \frac{M_1 \times V_1}{V_2}$$

### COMPUTE

Note: Even though molarity is moles per liter, you can use volumes in milliliters along with molarity whenever the units cancel.

$$M_2 = \frac{4.74 \text{ M} \times 50.00 \text{ mL}}{250.00 \text{ mL}} = 0.948 \text{ M}$$

### EVALUATE

Are the units correct?
**Yes; molarity (mol/L) was required.**

Is the number of significant figures correct?
**Yes; the number of significant figures is correct because the data had a minimum of three significant figures.**

Is the answer reasonable?
**Yes; the computation is the same as 4.74/5, which is a little less than 1.**

# Practice

**1.** Complete the table below by calculating the missing value in each row.

| | Molarity of stock solution | Volume of stock solution | Molarity of dilute solution | Volume of dilute solution | |
|---|---|---|---|---|---|
| **a.** | 0.500 M KBr | 20.00 mL | ? M KBr | 100.00 mL | ans: 0.100 M |
| **b.** | 1.00 M LiOH | ? mL | 0.075 M LiOH | 500.00 mL | ans: 38 mL |
| **c.** | ? M HI | 5.00 mL | 0.0493 M HI | 100.00 mL | ans: 0.986 M |
| **d.** | 12.0 M HCl | 0.250 L | 1.8 M HCl | ? L | ans: 1.7 L |
| **e.** | 7.44 M NH$_3$ | ? mL | 0.093 M NH$_3$ | 4.00 L | ans: 50. mL |

# Sample Problem 2

**What volume of water would you add to 15.00 mL of a 6.77 M solution of nitric acid in order to get a 1.50 M solution?**

## Solution

### ANALYZE

What is given in the problem?    **the molarity of the stock solution, the volume of stock solution, and the molarity of the diluted solution**

What are you asked to find?    **the volume of water to add to make the dilute solution**

| Items | Data |
|---|---|
| Concentration of the stock solution ($M_1$) | 6.77 M $HNO_3$ |
| Volume of stock solution used ($V_1$) | 15.00 mL |
| Molarity of the diluted solution ($M_2$) | 1.50 M $HNO_3$ |
| Volume of diluted solution ($V_2$) | ? mL |
| Volume of water to add | ? mL |

### PLAN

What steps are needed to calculate the amount of water to add to dilute a solution to the given molarity?

**Apply the principle that *volume₁ × molarity₁ = volume₂ × molarity₂*. Subtract the stock solution volume from the final volume to determine the amount of water to add.**

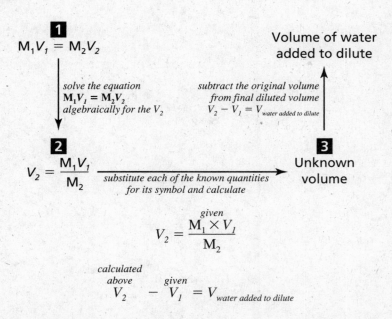

$$V_2 = \frac{\overset{given}{M_1 \times V_1}}{M_2}$$

$$\underset{above}{\overset{calculated}{V_2}} - \underset{given}{V_1} = V_{\text{water added to dilute}}$$

**| Problem Solving** *continued*

## COMPUTE

$$V_2 = \frac{6.77 \text{ M} \times 15.00 \text{ mL}}{1.50 \text{ M}} = 67.7 \text{ mL}$$

$$67.7 \text{ mL} - 15.00 \text{ mL} = 52.7 \text{ mL } H_2O$$

## EVALUATE

Are the units correct?

**Yes; volume of water was required.**

Is the number of significant figures correct?

**Yes; the number of significant figures is correct because the data had a minimum of three significant figures.**

Is the answer reasonable?

**Yes; the dilution was to a concentration of less than 1/4 of the original concentration. Thus, the volume of the diluted solution should be more than four times the original volume, 4 × 15 mL = 60 mL.**

# Practice

1. What volume of water would be added to 16.5 mL of a 0.0813 M solution of sodium borate in order to get a 0.0200 M solution? **ans: 50.6 mL $H_2O$**

## Additional Problems

1. What is the molarity of a solution of ammonium chloride prepared by diluting 50.00 mL of a 3.79 M $NH_4Cl$ solution to 2.00 L?

2. A student takes a sample of KOH solution and dilutes it with 100.00 mL of water. The student determines that the diluted solution is 0.046 M KOH, but has forgotten to record the volume of the original sample. The concentration of the original solution is 2.09 M. What was the volume of the original sample?

3. A chemist wants to prepare a stock solution of $H_2SO_4$ so that samples of 20.00 mL will produce a solution with a concentration of 0.50 M when added to 100.0 mL of water.

   a. What should the molarity of the stock solution be?

   b. If the chemist wants to prepare 5.00 L of the stock solution from concentrated $H_2SO_4$, which is 18.0 M, what volume of concentrated acid should be used?

   c. The density of 18.0 M $H_2SO_4$ is 1.84 g/mL. What mass of concentrated $H_2SO_4$ should be used to make the stock solution in (b)?

4. To what volume should 1.19 mL of an 8.00 M acetic acid solution be diluted in order to obtain a final solution that is 1.50 M?

5. What volume of a 5.75 M formic acid solution should be used to prepare 2.00 L of a 1.00 M formic acid solution?

6. A 25.00 mL sample of ammonium nitrate solution produces a 0.186 M solution when diluted with 50.00 mL of water. What is the molarity of the stock solution?

7. Given a solution of known percentage concentration by mass, a laboratory worker can often measure out a calculated mass of the solution in order to obtain a certain mass of solute. Sometimes, though, it is impractical to use the mass of a solution, especially with fuming solutions, such as concentrated HCl and concentrated $HNO_3$. Measuring these solutions by volume is much more practical. In order to determine the volume that should be measured, a worker would need to know the density of the solution. This information usually appears on the label of the solution bottle.

   a. Concentrated hydrochloric acid is 36% HCl by mass and has a density of 1.18 g/mL. What is the volume of 1.0 kg of this HCl solution? What volume contains 1.0 g of HCl? What volume contains 1.0 mol of HCl?

   b. The density of concentrated nitric acid is 1.42 g/mL, and its concentration is 71% $HNO_3$ by mass. What volume of concentrated $HNO_3$ would be needed to prepare 10.0 L of a 2.00 M solution of $HNO_3$?

   c. What volume of concentrated HCl solution would be needed to prepare 4.50 L of 3.0 M HCl? See (a) for data.

8. A 3.8 M solution of $FeSO_4$ solution is diluted to eight times its original volume. What is the molarity of the diluted solution?

**9.** A chemist prepares 480. mL of a 2.50 M solution of $K_2Cr_2O_7$ in water. A week later, the chemist wants to use the solution, but the stopper has been left off the flask and 39 mL of water has evaporated. What is the new molarity of the solution?

**10.** You must write out procedures for a group of lab technicians. One test they will perform requires 25.00 mL of a 1.22 M solution of acetic acid. You decide to use a 6.45 M acetic acid solution that you have on hand. What procedure should the technicians use in order to get the solution they need?

**11.** A chemical test has determined the concentration of a solution of an unknown substance to be 2.41 M. A 100.0 mL volume of the solution is evaporated to dryness, leaving 9.56 g of crystals of the unknown solute. Calculate the molar mass of the unknown substance.

**12.** Tincture of iodine can be prepared by dissolving 34 g of $I_2$ and 25 g of KI in 25 mL of distilled water and diluting the solution to 500. mL with ethanol. What is the molarity of $I_2$ in the solution?

**13.** Phosphoric acid is commonly supplied as an 85% solution. What mass of this solution would be required to prepare 600.0 mL of a 2.80 M phosphoric acid solution?

**14.** Commercially available concentrated sulfuric acid is 18.0 M $H_2SO_4$. What volume of concentrated $H_2SO_4$ would you use in order to make 3.00 L of a 4.0 M stock solution?

**15.** Describe how to prepare 1.00 L of a 0.495 M solution of urea, $NH_2CONH_2$, starting with a 3.07 M stock solution.

**16.** Honey is a solution consisting almost entirely of a mixture of the hexose sugars fructose and glucose; both sugars have the formula $C_6H_{12}O_6$, but they differ in molecular structure.

   **a.** A sample of honey is found to be 76.2% $C_6H_{12}O_6$ by mass. What is the molality of the hexose sugars in honey? Consider the sugars to be equivalent.

   **b.** The density of the honey sample is 1.42 g/mL. What mass of hexose sugars are in 1.00 L of honey? What is the molarity of the mixed hexose sugars in honey?

**17.** Industrial chemicals used in manufacturing are almost never pure, and the content of the material may vary from one batch to the next. For these reasons, a sample is taken from each shipment and sent to a laboratory, where its makeup is determined. This procedure is called assaying. Once the content of a material is known, engineers adjust the manufacturing process to account for the degree of purity of the starting chemicals.

   Suppose you have just received a shipment of sodium carbonate, $Na_2CO_3$. You weigh out 50.00 g of the material, dissolve it in water, and dilute the solution to 1.000 L. You remove 10.00 mL from the solution and dilute it to 50.00 mL. By measuring the amount of a second substance that reacts with

$Na_2CO_3$, you determine that the concentration of sodium carbonate in the diluted solution is 0.0890 M. Calculate the percentage of $Na_2CO_3$ in the original batch of material. The molar mass of $Na_2CO_3$ is 105.99 g. (Hint: Determine the number of moles in the original solution and convert to mass of $Na_2CO_3$.)

**18.** A student wants to prepare 0.600 L of a stock solution of copper(II) chloride so that 20.0 mL of the stock solution diluted by adding 130.0 mL of water will yield a 0.250 M solution. What mass of $CuCl_2$ should be used to make the stock solution?

**19.** You have a bottle containing a 2.15 M $BaCl_2$ solution. You must tell other students how to dilute this solution to get various volumes of a 0.65 M $BaCl_2$ solution. By what factor will you tell them to dilute the stock solution? In other words, when a student removes any volume, $V$, of the stock solution, how many times $V$ of water should be added to dilute to 0.65 M?

**20.** You have a bottle containing an 18.2% solution of strontium nitrate (density = 1.02 g/mL).

**a.** What mass of strontium nitrate is dissolved in 80.0 mL of this solution?

**b.** How many moles of strontium nitrate are dissolved in 80.0 mL of the solution?

**c.** If 80.0 mL of this solution is diluted with 420.0 mL of water, what is the molarity of the solution?

Skills Worksheet

# Problem Solving

## Colligative Properties

Colligative properties of solutions are properties that depend solely on the number of particles of solute in solution. In other words, these properties depend only on the concentration of solute particles, not on the identity of those particles. Colligative properties result from the interference of solute particles with the motion of solvent molecules.

Solute molecules can be either electrolytes or nonelectrolytes. When a nonelectrolyte dissolves, the molecule remains whole in the solution. Glucose and glycerol are examples of nonelectrolyte solutes.

Ionic solutes are electrolytes. When they dissolve, they dissociate into multiple particles, or ions. When magnesium chloride dissolves in water, it dissociates as follows.

$$MgCl_2(s) \xrightarrow{H_2O} Mg^{2+}(aq) + 2Cl^-(aq)$$

As you can see, when a mole of $MgCl_2$ completely dissociates in solution, it produces 1 mol of $Mg^{2+}$ ions and 2 mol of $Cl^-$ ions for a total of 3 mol of solute particles. Because colligative properties depend on the number of particles in solution, 1 mol of $MgCl_2$ in solution should have three times the effect of 1 mol of a nonelectrolyte solute, such as glucose.

Two important colligative properties are freezing-point depression and boiling-point elevation. A dissolved solute lowers the freezing point of the solution. The freezing point of a solution differs from that of the pure solvent according to the following equation, in which $\Delta t_f$ is the change in freezing point.

$$\Delta t_f = K_f m$$

$K_f$ is a constant that differs for each solvent. Because the freezing point of the solution is lower than that of the solvent alone, $K_f$ is a negative number. The symbol $m$ represents the molality (moles of solute per kilogram of solvent) of the solution.

Boiling-point elevation works in the same way. The equation to determine the change in boiling point is as follows.

$$\Delta t_b = K_b m$$

Like $K_f$, $K_b$ is a constant that differs for each solvent. But unlike $K_f$, $K_b$ is a positive number because the boiling point of the solution is higher than that of the solvent alone.

**General Plan for Solving Problems Involving Freezing-Point Depression and Boiling-Point Elevation**

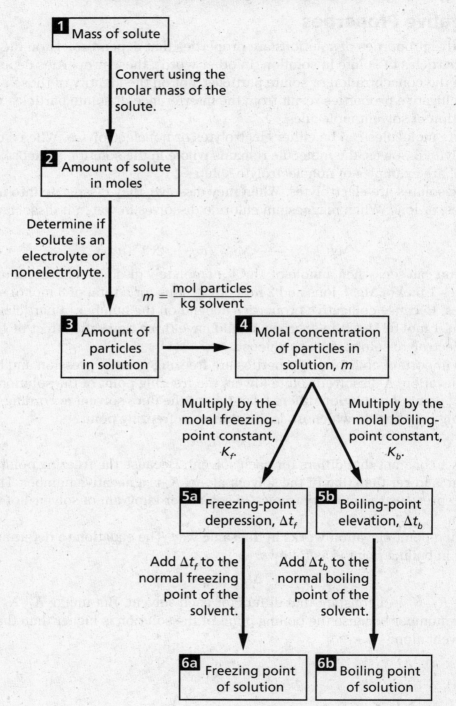

1 Mass of solute

Convert using the molar mass of the solute.

2 Amount of solute in moles

Determine if solute is an electrolyte or nonelectrolyte.

$$m = \frac{\text{mol particles}}{\text{kg solvent}}$$

3 Amount of particles in solution

4 Molal concentration of particles in solution, $m$

Multiply by the molal freezing-point constant, $K_f$.

Multiply by the molal boiling-point constant, $K_b$.

5a Freezing-point depression, $\Delta t_f$

5b Boiling-point elevation, $\Delta t_b$

Add $\Delta t_f$ to the normal freezing point of the solvent.

Add $\Delta t_b$ to the normal boiling point of the solvent.

6a Freezing point of solution

6b Boiling point of solution

**Problem Solving** *continued*

Table 1 lists freezing-point depression and boiling-point elevation constants for common solvents.

## TABLE 1

| Solvent | Normal f.p. | $K_f$ | Normal b.p. | $K_b$ |
|---|---|---|---|---|
| Acetic acid | 16.6°C | −3.90°C/$m$ | 117.9°C | 3.07°C/$m$ |
| Camphor | 178.8°C | −39.7°C/$m$ | 207.4°C | 5.61°C/$m$ |
| Ether | −116.3°C | −1.79°C/$m$ | 34.6°C | 2.02°C/$m$ |
| Naphthalene | 80.2°C | −6.94°C/$m$ | 217.7°C | 5.80°C/$m$ |
| Phenol | 40.9°C | −7.40°C/$m$ | 181.8°C | 3.60°C/$m$ |
| Water | 0.00°C | −1.86°C/$m$ | 100.0°C | 0.51°C/$m$ |

# Sample Problem 1

**What is the freezing point of a solution of 210.0 g of glycerol, HOCH₂CHOHCH₂OH, dissolved in 350. g of water?**

## Solution

### ANALYZE

What is given in the problem?  **the formula and mass of solute, and the mass of water used**

What are you asked to find?  **the freezing point of the solution**

| Items | Data |
|---|---|
| Identity of solute | glycerol, $HOCH_2CHOHCH_2OH$ |
| Particles per mole of solute | 1 mol |
| Identity of solvent | water |
| Freezing point of solvent | 0.00°C |
| Mass of solvent | 350. g |
| Mass of solute | 210.0 g |
| Molar mass of solute* | 92.11 g/mol |
| Molal concentration of solute particles | $?\ m$ |
| Molal freezing-point constant for water | $-1.86°C/m$ |
| Freezing-point depression | ?°C |
| Freezing point of solution | ?°C |

*determined from the periodic table

### PLAN

What steps are needed to calculate the freezing point of the solution?
**Use the molar mass of the solute to determine the amount of solute. Then apply the mass of solvent to calculate the molality of the solution. From the molality, use the molal freezing-point constant for water to calculate the number of degrees the freezing point is lowered. Add this negative value to the normal freezing point.**

**Problem Solving** *continued*

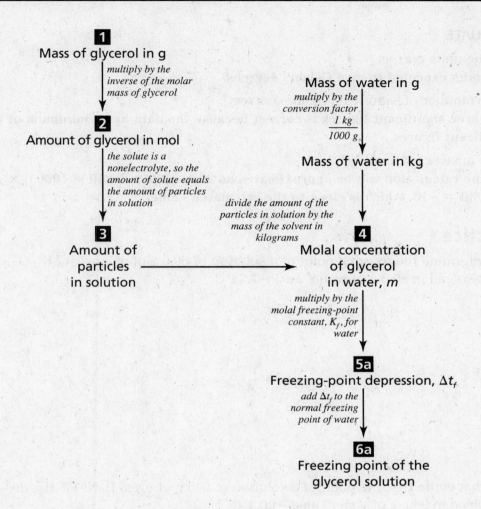

**1**
Mass of glycerol in g

*multiply by the inverse of the molar mass of glycerol*

**2**
Amount of glycerol in mol

*the solute is a nonelectrolyte, so the amount of solute equals the amount of particles in solution*

**3**
Amount of particles in solution

Mass of water in g

*multiply by the conversion factor*
$$\frac{1\ kg}{1000\ g}$$

Mass of water in kg

*divide the amount of the particles in solution by the mass of the solvent in kilograms*

**4**
Molal concentration of glycerol in water, *m*

*multiply by the molal freezing-point constant, $K_f$, for water*

**5a**
Freezing-point depression, $\Delta t_f$

*add $\Delta t_f$ to the normal freezing point of water*

**6a**
Freezing point of the glycerol solution

$$\overset{given}{g\ H_2O} \times \frac{1\ kg}{1000\ g} = kg\ H_2O$$

$$\overset{given}{g\ glycerol} \times \underset{molar\ mass\ of\ glycerol}{\frac{1\ mol\ glycerol}{92.11\ g\ glycerol}} \times \overset{calculated\ above}{\frac{1}{kg\ H_2O}} \times \overset{freezing-point\ depression\ constant}{\frac{-1.86°C}{mol/kg}} = \Delta t_f$$

$$\underset{\substack{freezing\ point \\ of\ H_2O}}{0.00°C} + \underset{\substack{calculated \\ above}}{\Delta t_f} = t_f$$

**COMPUTE**

$$350.\ g\ H_2O \times \frac{1\ kg}{1000\ g} = 0.350\ kg\ H_2O$$

$$210.0\ g\ glycerol \times \frac{1\ mol\ glycerol}{92.11\ g\ glycerol} \times \frac{1}{0.350\ kg\ H_2O} \times \frac{-1.86°C}{mol/kg} = -12.1°C$$

$$0.00°C + (-12.1°C) = -12.1°C$$

## EVALUATE

Are the units correct?

**Yes; units canceled to give Celsius degrees.**

Is the number of significant figures correct?

**Yes; three significant figures is correct because the data had a minimum of three significant figures.**

Is the answer reasonable?

**Yes; the calculation can be approximated as $200 \div [90 \times 3(350 \div 1000)] \times -2 \approx -400/30 = -13$, which is close to the calculated value.**

# Practice

1. Determine the freezing point of a solution of 60.0 g of glucose, $C_6H_{12}O_6$, dissolved in 80.0 g of water. **ans: −7.74°C**

2. What is the freezing point of a solution of 645 g of urea, $H_2NCONH_2$, dissolved in 980. g of water? **ans: −20.4°C**

**Problem Solving** *continued*

# Sample Problem 2

What is the boiling point of a solution containing 34.3 g of the ionic
compound magnesium nitrate dissolved in 0.107 kg of water?

# Solution

## ANALYZE

What is given in the problem?  **the formula and mass of solute, and the mass of water used**

What are you asked to find?  **the boiling point of the solution**

| Items | Data |
|---|---|
| Identity of solute | magnesium nitrate |
| Equation for the dissociation of the solute | $Mg(NO_3)_2 \rightarrow Mg^{2+} + 2NO_3^-$ |
| Amount of ions per mole of solute | 3 mol |
| Identity of solvent | water |
| Boiling point of solvent | 100.0°C |
| Mass of solvent | 0.107 kg $H_2O$ |
| Mass of solute | 34.3 g |
| Molar mass of solute | 148.32 g/mol |
| Molal concentration of solute particles | $? \; m$ |
| Molal boiling-point constant for solvent | 0.51°C/$m$ |
| Boiling-point depression | ?°C |
| Boiling point of solution | ?°C |

## PLAN

What steps are needed to calculate the boiling point of the solution?
**Use the molar mass to calculate the amount of solute in moles. Multiply the
amount of solute by the number of moles of ions produced per mole of solute.
Use the amount of ions with the mass of solvent to compute the molality of
particles in solution. Use this effective molality to determine the boiling-point
elevation and the boiling point of the solution.**

**▌Problem Solving** *continued*

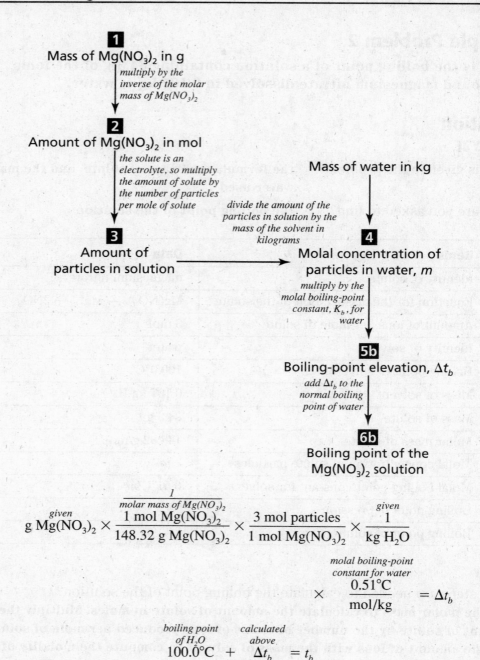

**1**
Mass of Mg(NO$_3$)$_2$ in g

*multiply by the inverse of the molar mass of Mg(NO$_3$)$_2$*

**2**
Amount of Mg(NO$_3$)$_2$ in mol

*the solute is an electrolyte, so multiply the amount of solute by the number of particles per mole of solute*

Mass of water in kg

*divide the amount of the particles in solution by the mass of the solvent in kilograms*

**3**
Amount of
particles in solution →

**4**
Molal concentration of
particles in water, *m*

*multiply by the molal boiling-point constant, K$_b$, for water*

**5b**
Boiling-point elevation, Δt$_b$

*add Δt$_b$ to the normal boiling point of water*

**6b**
Boiling point of the
Mg(NO$_3$)$_2$ solution

$$\text{g Mg(NO}_3)_2 \overset{given}{} \times \frac{\overset{\frac{1}{molar\ mass\ of\ Mg(NO_3)_2}}{1\ \text{mol Mg(NO}_3)_2}}{148.32\ \text{g Mg(NO}_3)_2} \times \frac{3\ \text{mol particles}}{1\ \text{mol Mg(NO}_3)_2} \times \overset{given}{\frac{1}{\text{kg H}_2\text{O}}}$$

$$\times \overset{\substack{molal\ boiling\text{-}point \\ constant\ for\ water}}{\frac{0.51°\text{C}}{\text{mol/kg}}} = \Delta t_b$$

$$\underset{\substack{boiling\ point \\ of\ H_2O}}{100.0°\text{C}} + \underset{\substack{calculated \\ above}}{\Delta t_b} = t_b$$

**COMPUTE**

$$34.3\ \text{g Mg(NO}_3)_2 \times \frac{1\ \text{mol Mg(NO}_3)_2}{148.32\ \text{g Mg(NO}_3)_2} \times \frac{3\ \text{mol particles}}{1\ \text{mol Mg(NO}_3)_2} \times \frac{1}{0.107\ \text{kg H}_2\text{O}}$$

$$\times \frac{0.51°\text{C}}{\text{mol/kg}} = 3.31°\text{C}$$

$$100.0°\text{C} + 3.31°\text{C} = 103.3°\text{C}$$

**Problem Solving** *continued*

**EVALUATE**

Are the units correct?

**Yes; units canceled to give Celsius degrees.**

Is the number of significant figures correct?

**Yes; the number of significant figures is correct because the boiling point of water was given to one decimal place.**

Is the answer reasonable?

**Yes; the calculation can be approximated as [(35 × 3)/150] × 5 = (7/10) × 5 = 3.5, which is close to the calculated value for the boiling-point elevation.**

# Practice

**1.** What is the expected boiling point of a brine solution containing 30.00 g of KBr dissolved in 100.00 g of water? **ans: 102.6°C**

**2.** What is the expected boiling point of a $CaCl_2$ solution containing 385 g of $CaCl_2$ dissolved in $1.230 \times 10^3$ g of water? **ans: 104.3°C**

**❙** Problem Solving *continued*

## Sample Problem 3

A solution of 3.39 g of an unknown compound in 10.00 g of water has a freezing point of −7.31°C. The solution does not conduct electricity. What is the molar mass of the compound?

## Solution

### ANALYZE

What is given in the problem?    the freezing point of the solution, the mass of the dissolved compound, the mass of solvent, and the fact that the solution does not conduct electricity

What are you asked to find?    the molar mass of the unknown compound

| Items | Data |
|-------|------|
| Mass of solute | 3.39 g |
| Molar mass of solute | ? g/mol |
| Identity of solvent | water |
| Freezing point of solvent | 0.00°C |
| Mass of solvent | 10.00 g |
| Molal freezing-point constant for solvent | −1.86°C/$m$ |
| Freezing-point depression | ?°C |
| Freezing point of solution | −7.31°C |
| Molal concentration of solute particles | ? $m$ |

### PLAN

What steps are needed to calculate the molar mass of the unknown solute?
**Determine the molality of the solution from the freezing-point depression. Use the molality and the solute and solvent masses to calculate the solute molar mass.**

**▌Problem Solving** *continued*

**6a**
Freezing point of the solution

*subtract the normal freezing point of water*
↓

**5a**
Freezing-point depression, $\Delta t_f$

*multiply by the inverse of the molal freezing-point constant, $K_f$, for water*
↓

**4**
Molal concentration of particles in water, $m$ ────────→

Molar mass of solute
↑
*divide the mass of the solute by the amount of solute in moles*

**2**
Amount of solute in mol
↑
*the solute is a nonelectrolyte, so the amount of solute equals the amount of particles in solution*

**3**
Amount of particles in solution
↑

*multiply the molal concentration by the mass of the water*

Mass of water in g ────────→ Mass of water in kg

*multiply by the conversion factor $\frac{1\ kg}{1000\ g}$*

$$\underset{given}{t_f} - \underset{\substack{freezing\ point \\ of\ water}}{0.00°C} = \Delta t_f$$

$$\underset{given}{g\ H_2O} \times \frac{1\ kg}{1000\ g} = kg\ H_2O$$

$$\underset{\substack{calculated \\ above}}{\Delta t_f} \times \underset{\substack{\frac{1}{molal\ freezing-point} \\ constant\ for\ water}}{\frac{mol/kg}{-1.86°C}} \times \underset{\substack{calculated \\ above}}{kg\ H_2O} = mol\ solute$$

$$\frac{\underset{given}{g\ solute}}{\underset{calculated\ above}{mol\ solute}} = molar\ mass\ of\ solute$$

**COMPUTE**

$$-7.31°C - 0.00°C = -7.31°C$$

$$10.00\ \cancel{g}\ H_2O \times \frac{1\ kg}{1000\ \cancel{g}} = 0.010\ 00\ kg\ H_2O$$

$$-7.31\cancel{°C} \times \frac{mol/kg}{-1.86\cancel{°C}} \times 0.010\ 00\ \cancel{kg\ H_2O} = 0.039\ 30\ mol\ solute$$

$$\frac{3.39\ g\ solute}{0.039\ 30\ mol\ solute} = 86.3\ g/mol$$

## EVALUATE

Are the units correct?

**Yes; molar mass has units of g/mol.**

Is the number of significant figures correct?

**Yes; the number of significant figures is correct because the data had a minimum of three significant figures.**

Is the answer reasonable?

**Yes; the calculation can be approximated as (4/1) × (1/100) = 0.04, which is close to the value of 0.0393 for the amount of solute.**

## Practice

**1.** A solution of 0.827 g of an unknown non-electrolyte compound in 2.500 g of water has a freezing point of −10.18°C. Calculate the molar mass of the compound. **ans: 60.4 g/mol**

**2.** A 0.171 g sample of an unknown organic compound is dissolved in ether. The solution has a total mass of 2.470 g. The boiling point of the solution is found to be 36.43°C. What is the molar mass of the organic compound?
**ans: 82.1 g/mol**

**▌Problem Solving** *continued*

# Additional Problems

**In each of the following problems, assume that the solute is a nonelectrolyte unless otherwise stated.**

**1.** Calculate the freezing point and boiling point of a solution of 383 g of glucose dissolved in 400. g of water.

**2.** Determine the boiling point of a solution of 72.4 g of glycerol dissolved in 122.5 g of water.

**3.** What is the boiling point of a solution of 30.20 g of ethylene glycol, $HOCH_2CH_2OH$, in 88.40 g of phenol?

**4.** What mass of ethanol, $CH_3CH_2OH$, should be dissolved in 450. g of water to obtain a freezing point of $-4.5°C$?

**5.** Calculate the molar mass of a nonelectrolyte that lowers the freezing point of 25.00 g of water to $-3.9°C$ when 4.27 g of the substance is dissolved in the water.

**6.** What is the freezing point of a solution of 1.17 g of 1-naphthol, $C_{10}H_8O$, dissolved in 2.00 mL of benzene at 20°C? The density of benzene at 20°C is 0.876 g/mL. $K_f$ for benzene is $-5.12°C/m$, and benzene's normal freezing point is 5.53°C.

**7.** The boiling point of a solution containing 10.44 g of an unknown nonelectrolyte in 50.00 g of acetic acid is 159.2°C. What is the molar mass of the solute?

**8.** A 0.0355 g sample of an unknown molecular compound is dissolved in 1.000 g of liquid camphor at 200.0°C. Upon cooling, the camphor freezes at 157.7°C. Calculate the molar mass of the unknown compound.

**9.** Determine the boiling point of a solution of 22.5 g of fructose, $C_6H_{12}O_6$, in 294 g of phenol.

**10.** Ethylene glycol, $HOCH_2CH_2OH$, is effective as an antifreeze, but it also raises the boiling temperature of automobile coolant, which helps prevent loss of coolant when the weather is hot.

   **a.** What is the freezing point of a 50.0% solution of ethylene glycol in water?

   **b.** What is the boiling point of the same 50.0% solution?

**11.** The value of $K_f$ for cyclohexane is $-20.0°C/m$, and its normal freezing point is 6.6°C. A mass of 1.604 g of a waxy solid dissolved in 10.000 g of cyclohexane results in a freezing point of $-4.4°C$. Calculate the molar mass of the solid.

**12.** What is the expected freezing point of an aqueous solution of 2.62 kg of nitric acid, $HNO_3$, in a solution with a total mass of 5.91 kg? Assume that the nitric acid is completely ionized.

**Problem Solving** *continued*

13. An unknown organic compound is mixed with 0.5190 g of naphthalene crystals to give a mixture having a total mass of 0.5959 g. The mixture is heated until the naphthalene melts and the unknown substance dissolves. Upon cooling, the solution freezes at a temperature of 74.8°C. What is the molar mass of the unknown compound?

14. What is the boiling point of a solution of 8.69 g of the electrolyte sodium acetate, $NaCH_3COO$, dissolved in 15.00 g of water?

15. What is the expected freezing point of a solution of 110.5 g of $H_2SO_4$ in 225 g of water? Assume sulfuric acid completely dissociates in water.

16. A compound called pyrene has the empirical formula $C_8H_5$. When 4.04 g of pyrene is dissolved in 10.00 g of benzene, the boiling point of the solution is 85.1°C. Calculate the molar mass of pyrene and determine its molecular formula. The molal boiling-point constant for benzene is 2.53°C/$m$. Its normal boiling point is 80.1°C.

17. What mass of $CaCl_2$, when dissolved in 100.00 g of water, gives an expected freezing point of −5.0°C; $CaCl_2$ is ionic? What mass of glucose would give the same result?

18. A compound has the empirical formula $CH_2O$. When 0.0866 g is dissolved in 1.000 g of ether, the solution's boiling point is 36.5°C. Determine the molecular formula of this substance.

19. What is the freezing point of a 28.6% (by mass) aqueous solution of HCl? Assume the HCl is 100% ionized.

20. What mass of ethylene glycol, $HOCH_2CH_2OH$, must be dissolved in 4.510 kg of water to result in a freezing point of −18.0°C? What is the boiling point of the same solution?

21. A water solution containing 2.00 g of an unknown molecular substance dissolved in 10.00 g of water has a freezing point of −4.0°C.

    **a.** Calculate the molality of the solution.

    **b.** When 2.00 g of the substance is dissolved in acetone instead of in water, the boiling point of the solution is 58.9°C. The normal boiling point of acetone is 56.00°C, and its $K_b$ is 1.71°C/$m$. Calculate the molality of the solution from this data.

22. A chemist wants to prepare a solution with a freezing point of −22.0°C and has 100.00 g of glycerol on hand. What mass of water should the chemist mix with the glycerol?

23. An unknown carbohydrate compound has the empirical formula $CH_2O$. A solution consisting of 0.515 g of the carbohydrate dissolved in 1.717 g of acetic acid freezes at 8.8°C. What is the molar mass of the carbohydrate? What is its molecular formula?

24. An unknown organic compound has the empirical formula $C_2H_2O$. A solution of 3.775 g of the unknown compound dissolved in 12.00 g of water is cooled until it freezes at a temperature of −4.72°C. Determine the molar mass and the molecular formula of the compound.

# Problem Solving

## Equilibrium

Not all processes in nature proceed to completion. In fact, most changes hover somewhere between the initial state and what would be the final state. Compare a light switch and a dimmer. If the mechanical switch is working properly, it can be stable in only two positions: on or off. Either current flows or it doesn't. With a dimmer you can regulate the flow of current so that it stays somewhere between fully on and fully off. If you've used mechanical balances, you know that to weigh an object accurately you must adjust the masses so that the pointer hovers in the middle of its range. The balance is in a state of equilibrium.

Most chemical reactions also reach a state of equilibrium between no reaction at all and the complete reaction to form products. Equilibrium states occur when reactions are reversible, that is, when products react to re-form the original reactants. When the products re-form reactants at the same rate as the reactants form the products, then the equilibrium point of the reaction has been reached.

The progress of a reaction is gauged by measuring the concentrations in moles per liter of reactants and products. At the equilibrium point, these concentrations stop changing.

$$A + 2B \rightleftharpoons 2C$$

The equation above represents a reaction in which 1 mol of A reacts with 2 mol of B to produce 2 mol of C. As C is formed, it breaks down to re-form reactants. The extent of the reaction at equilibrium is indicated by the equilibrium constant, $K_{eq}$.

$$\frac{K_{eq} = [C]^2}{[A][B]}$$

As you can see, the concentration of each reaction component is raised to the power of its coefficient in the balanced equation. These concentration terms are arranged in a fraction with products in the numerator and reactants in the denominator. Pure substances (substances that appear in the chemical equation as solids or pure liquids) are not included in the equilibrium expressions because their concentrations are meaningless.

Problems involving chemical equilibrium will ask you to solve for either the equilibrium constant, $K_{eq}$, given the concentrations of all of the reaction components, or the concentration of one of the reaction components, given $K_{eq}$.

## General Plan for Solving Equilibrium Problems

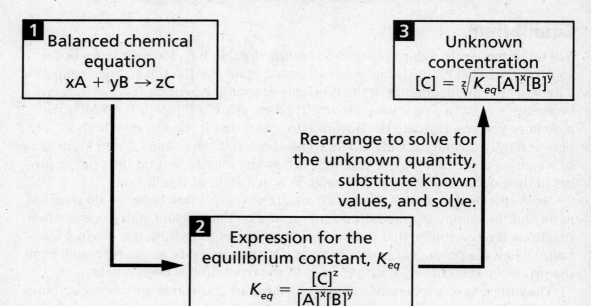

**1** Balanced chemical equation
$$xA + yB \rightarrow zC$$

**3** Unknown concentration
$$[C] = \sqrt[z]{K_{eq}[A]^x[B]^y}$$

Rearrange to solve for the unknown quantity, substitute known values, and solve.

**2** Expression for the equilibrium constant, $K_{eq}$
$$K_{eq} = \frac{[C]^z}{[A]^x[B]^y}$$

# Sample Problem 1

The following equation represents the reversible decomposition of $PCl_5$.

$$PCl_5(g) \rightleftarrows PCl_3(g) + Cl_2(g)$$

At 250°C, the equilibrium concentrations of the substances are as follows:

$[PCl_5] = 1.271$ M

$[PCl_3] = 0.229$ M

$[Cl_2] = 0.229$ M

What is the value of the equilibrium constant, $K_{eq}$, for this reaction?

# Solution

## ANALYZE

What is given in the problem?     **the equilibrium concentrations of the products and reactant at 250°C**

What are you asked to find?     **the value of the equilibrium constant for the reaction**

| Items | Data |
|---|---|
| Molar concentration of $PCl_5$ at equilibrium | 1.271 |
| Molar concentration of $PCl_3$ at equilibrium | 0.229 M |
| Molar concentration of $Cl_2$ at equilibrium | 0.229 M |
| Equilibrium constant $K_{eq}$ | ? |

## PLAN

What steps are needed to calculate the equilibrium constant for the given reaction?
**Set up the equilibrium expression for the reaction using the coefficients as exponents. Substitute the concentration values, and calculate $K_{eq}$.**

$$\boxed{1}$$
$$PCl_5(g) \rightleftharpoons PCl_3(g) + Cl_2(g) \longrightarrow \boxed{2} \quad K_{eq} = \frac{[PCl_3][Cl_2]}{[PCl_5]}$$

Note that since all coefficients have the value 1, there is no need to write in the exponent.

$$K_{eq} = \frac{\overset{given \quad given}{[PCl_3][Cl_2]}}{\underset{given}{[PCl_5]}}$$

**COMPUTE**

$$K_{eq} = \frac{[0.229][0.229]}{[1.271]} = 0.0413$$

**EVALUATE**

Are the units correct?

**Yes; the equilibrium constant has no units.**

Is the number of significant figures correct?

**Yes; the number of significant figures is correct because data values were given to a minimum of three significant figures.**

Is the answer reasonable?

**Yes; the calculation may be approximated as (0.2 × 0.2)/1. This approximation gives a result of 0.04, which is close to the calculated value.**

## Practice

1. Calculate the equilibrium constants for the following hypothetical reactions. Assume that all components of the reactions are gaseous.

   **a.** $A \rightleftharpoons C + D$

   At equilibrium, the concentration of A is $2.24 \times 10^{-2}$ M and the concentrations of both C and D are $6.41 \times 10^{-3}$ M. **ans: $K_{eq} = 1.83 \times 10^{-3}$**

   **b.** $A + B \rightleftharpoons C + D$

   At equilibrium, the concentrations of both A and B are $3.23 \times 10^{-5}$ M and the concentrations of both C and D are $1.27 \times 10^{-2}$ M. **ans: $K_{eq} = 1.55 \times 10^{5}$**

**c.** $A + B \rightleftarrows 2C$

At equilibrium, the concentrations of both A and B are $7.02 \times 10^{-3}$ M and the concentration of C is $2.16 \times 10^{-2}$ M. **ans: $K_{eq} = 9.47$**

**d.** $2A \rightleftarrows 2C + D$

At equilibrium, the concentration of A is $6.59 \times 10^{-4}$ M. The concentration of C is $4.06 \times 10^{-3}$ M, and the concentration of D is $2.03 \times 10^{-3}$ M. **ans: $K_{eq} = 7.71 \times 10^{-2}$**

**e.** $A + B \rightleftarrows C + D + E$

At equilibrium, the concentrations of both A and B are $3.73 \times 10^{-4}$ M and the concentrations of C, D, and E are $9.35 \times 10^{-4}$ M. **ans: $K_{eq} = 5.88 \times 10^{-3}$**

**f.** $2A + B \rightleftarrows 2C$

At equilibrium, the concentration of A is $5.50 \times 10^{-3}$ M, the concentration of B is $2.25 \times 10^{-3}$ M, and the concentration of C is $1.02 \times 10^{-2}$ M. **ans: $K_{eq} = 1.53 \times 10^{3}$**

## Sample Problem 2

**The following equilibrium reaction is used in the manufacture of methanol. The equilibrium constant at 400 K for the reaction is 1.609.**

$$CO(g) + 2H_2(g) \rightleftarrows CH_3OH(g)$$

**At equilibrium, the mixture in the reaction vessel has a concentration of 0.818 M of $CH_3OH$ and 1.402 M of CO. Calculate the concentration of $H_2$ in the vessel.**

## Solution

### ANALYZE

What is given in the problem?      the equilibrium concentrations of CO and
                                   $CH_3OH$, and the equilibrium constant at 400 K

What are you asked to find?        the equilibrium concentration of $H_2$ in the vessel

| Items | Data |
|-------|------|
| Molar concentration of CO at equilibrium | 1.402 M |
| Molar concentration of $H_2$ at equilibrium | ? M |
| Molar concentration of $CH_3OH$ at equilibrium | 0.818 M |
| Equilibrium constant $K_{eq}$ | 0.609 M |

### PLAN

What steps are needed to calculate the concentration of $H_2$?

Set up the equilibrium expression for the reaction using coefficients as exponents. Rearrange the expression to solve for [$H_2$]. Substitute known values for [$CO_2$], [$CH_3OH$], and $K_{eq}$, and solve for [$H_2$].

**1**
$$CO(g) + 2H_2(g) \rightleftarrows CH_3OH(g)$$

**2**
$$K_{eq} = \frac{[CH_3OH]}{[CO][H_2]^2}$$

*rearrange to solve for [$H_2$], substitute known values, and solve*

**3**
$$[H_2] = \sqrt{\frac{[CH_3OH]}{K_{eq} \times [CO]}}$$

| **Problem Solving** *continued*

## COMPUTE

$$[H_2] = \sqrt{\frac{[0.818]}{1.609 \times [1.402]}} = 0.602 \text{ M}$$

## EVALUATE

Are the units correct?
**Yes; concentrations are in moles per liter.**

Is the number of significant figures correct?
**Yes; the number of significant figures is correct because data values were given to a minimum of three significant figures.**

Is the answer reasonable?
**Yes; the calculation may be approximated as $[1 \div (1.5 \times 1.5)]^{1/2} = 0.67$, which is close to the calculated value.**

# Practice

**1.** Calculate the concentration of product D in the following hypothetical reaction:

$$2A(g) \rightleftarrows 2C(g) + D(g)$$

At equilibrium, the concentration of A is $1.88 \times 10^{-1}$ M, the concentration of C is 6.56 M, and the equilibrium constant is $2.403 \times 10^2$. **ans: 0.197 M**

**2.** At a temperature of 700 K, the equilibrium constant is $3.164 \times 10^3$ for the following reaction system for the hydrogenation of ethene, $C_2H_4$, to ethane, $C_2H_6$.

$$C_2H_4(g) + H_2(g) \rightleftarrows C_2H_6(g)$$

What will be the equilibrium concentration of ethene if the concentration of $H_2$ is 0.0619 M and the concentration of $C_2H_6$ is 1.055 M? **ans: $5.39 \times 10^{-3}$ M**

# Additional Problems—Equilibrium

1. Using the reaction A + 2B $\rightleftarrows$ C + 2D, determine $K_{eq}$ if the following equilibrium concentrations are found. All components are gases.

   [A] = 0.0567 M

   [B] = 0.1171 M

   [C] = 0.000 3378 M

   [D] = 0.000 6756 M

2. In the reaction 2A $\rightleftarrows$ 2C + 2D, determine $K_{eq}$ when the following equilibrium concentrations are found. All components are gases.

   [A] = 0.1077 M

   [C] = 0.000 4104 M

   [D] = 0.000 4104 M

3. Calculate the equilibrium constant for the following reaction. Note the phases of the components.

$$2A(g) + B(s) \rightleftarrows C(g) + D(g)$$

   The equilibrium concentrations of the components are

   [A] = 0.0922 M

   [C] = $4.11 \times 10^{-4}$ M

   [D] = $8.22 \times 10^{-4}$ M

4. The equilibrium constant of the following reaction for the decomposition of phosgene at 25°C is $4.282 \times 10^{-2}$.

$$COCl_2(g) \rightleftarrows CO(g) + Cl_2(g)$$

   **a.** What is the concentration of $COCl_2$ when the concentrations of both CO and $Cl_2$ are $5.90 \times 10^{-3}$ M?

   **b.** When the equilibrium concentration of $COCl_2$ is 0.003 70 M, what are the concentrations of CO and $Cl_2$? Assume the concentrations are equal.

5. Consider the following hypothetical reaction.

$$A(g) + B(s) \rightleftarrows C(g) + D(s)$$

   **a.** If $K_{eq} = 1$ for this reaction at 500 K, what can you say about the concentrations of A and C at equilibrium?

   **b.** If raising the temperature of the reaction results in an equilibrium with a higher concentration of C than A, how will the value of $K_{eq}$ change?

**6.** The following reaction occurs when steam is passed over hot carbon. The mixture of gases it generates is called water gas and is useful as an industrial fuel and as a source of hydrogen for the production of ammonia

$$C(s) + H_2O(g) \rightleftharpoons CO(g) + H_2(g)$$

The equilibrium constant for this reaction is $4.251 \times 10^{-2}$ at 800 K. If the equilibrium concentration of $H_2O(g)$ is 0.1990 M, what concentrations of CO and $H_2$ would you expect to find?

**7.** When nitrogen monoxide gas comes in contact with air, it oxidizes to the brown gas nitrogen dioxide according to the following equation:

$$2NO(g) + O_2(g) \rightleftharpoons 2NO_2(g)$$

**a.** The equilibrium constant for this reaction at 500 K is $1.671 \times 10^4$. What concentration of $NO_2$ is present at equilibrium if $[NO] = 6.200 \times 10^{-2}$ M and $[O_2] = 8.305 \times 10^{-3}$ M?

**b.** At 1000 K, the equilibrium constant, $K_{eq}$, for the same reaction is $1.315 \times 10^{-2}$. What will be the concentration of $NO_2$ at 1000 K given the same concentrations of NO and $O_2$ as were in (a)?

**8.** Consider the following hypothetical reaction, for which $K_{eq} = 1$ at 300 K:

$$A(g) + B(g) \rightleftharpoons 2C(g)$$

**a.** If the reaction begins with equal concentrations of A and B and a zero concentration of C, what can you say about the relative concentrations of the components at equilibrium?

**b.** Additional C is introduced at equilibrium, and the temperature remains constant. When equilibrium is restored, how will the concentrations of all components have changed? How will $K_{eq}$ have changed?

**9.** The equilibrium constant for the following reaction of hydrogen gas and bromine gas at 25°C is $5.628 \times 10^{18}$.

$$H_2(g) + Br_2(g) \rightleftharpoons 2HBr(g)$$

**a.** Write the equilibrium expression for this reaction.

**b.** Assume that equimolar amounts of $H_2$ and $Br_2$ were present at the beginning. Calculate the equilibrium concentration of $H_2$ if the concentration of HBr is 0.500 M.

**c.** If equal amounts of $H_2$ and $Br_2$ react, which reaction component will be present in the greatest concentration at equilibrium? Explain your reasoning.

**10.** The following reaction reaches an equilibrium state:

$$N_2F_4(g) \rightleftarrows 2NF_2(g)$$

At equilibrium at 25°C the concentration of $N_2F_4$ is found to be 0.9989 M and the concentration of $NF_2$ is $1.131 \times 10^{-3}$ M. Calculate the equilibrium constant of the reaction.

**11.** The equilibrium between dinitrogen tetroxide and nitrogen dioxide is represented by the following equation:

$$N_2O_4(g) \rightleftarrows NO_2(g)$$

A student places a mixture of the two gases into a closed gas tube and allows the reaction to reach equilibrium at 25°C. At equilibrium, the concentration of $N_2O_4$ is found to be $5.95 \times 10^{-1}$ M and the concentration of $NO_2$ is found to be $5.24 \times 10^{-2}$ M. What is the equilibrium constant of the reaction?

**12.** Consider the following equilibrium system:

$$NaCN(s) + HCl(g) \rightleftarrows HCN(g) + NaCl(s)$$

**a.** Write a complete expression for the equilibrium constant of this system.

**b.** The $K_{eq}$ for this reaction is $2.405 \times 10^6$. What is the concentration of HCl remaining when the concentration of HCN is 0.8959 M?

**13.** The following reaction is used in the industrial production of hydrogen gas:

$$CH_4(g) + H_2O(g) \rightleftarrows CO(g) + 3H_2(g)$$

The equilibrium constant of this reaction at 298 K (25°C) is $3.896 \times 10^{-27}$, but at 1100 K the constant is $3.112 \times 10^2$.

**a.** What do these equilibrium constants tell you about the progress of the reaction at the two temperatures?

**b.** Suppose the reaction mixture is sampled at 1100 K and found to contain 1.56 M of hydrogen, $3.70 \times 10^{-2}$ M of methane, and $8.27 \times 10^{-1}$ M of gaseous $H_2O$. What concentration of carbon monoxide would you expect to find?

**14.** Dinitrogen tetroxide, $N_2O_4$, is soluble in cyclohexane, a common nonpolar solvent. While in solution, $N_2O_4$ can break down into $NO_2$ according to the following equation:

$$N_2O_4(cyclohexane) \rightleftarrows NO_2(cyclohexane)$$

At 20°C, the following concentrations were observed for this equilibrium reaction:

$[N_2O_4] = 2.55 \times 10^{-3}$ M

$[NO_2] = 10.4 \times 10^{-3}$ M

What is the value of the equilibrium constant for this reaction? Note, the chemical equation must be balanced first.

**Problem Solving** *continued*

**15.** The reaction given in item 14 also occurs when the dinitrogen tetroxide and nitrogen dioxide are dissolved in carbon tetrachloride, $CCl_4$, another nonpolar solvent.

$$N_2O_4(CCl_4) \rightleftharpoons NO_2(CCl_4)$$

The following experimental data were obtained at 20°C:

$[N_2O_4] = 2.67 \times 10^{-3}$ M

$[NO_2] = 10.2 \times 10^{-3}$ M

Calculate the value of the equilibrium constant for this reaction occurring in carbon tetrachloride.

# Problem Solving

## Equilibrium of Salts, $K_{sp}$

When you try to dissolve a solid substance in water, you expect the solid form to disappear, forming ions or molecules in solution. Most substances, however, are only slightly soluble in water. For example, when you stir silver chloride in water, you may think none of the solid dissolves. Does this mean that some of the solid dissolves, forms a saturated solution, and, after that, experiences no further change between the solid and solution phases? It is true to say that there is no further change in the amount of substance in either the solution or the solid phase, but to say, that no further change occurs is inaccurate. An equilibrium exists between the silver chloride and its dissolved ions, and a state of equilibrium is a dynamic state. The following chemical equation shows this equilibrium.

$$AgCl(s) \rightleftarrows Ag^+(aq) + Cl^-(aq)$$

Like the other examples of equilibria that you have studied, the extent to which this solubility equilibrium proceeds toward the products (the ions in solution) is indicated by an equilibrium constant. When an equilibrium constant is written for a solubility equilibrium, it is called a solubility product constant and is symbolized as $K_{sp}$.

In solubility equilibrium problems, the reactants are pure substances, and pure substances are never included in an equilibrium expression. That means that you will not have anything in the denominator in the expression for the solubility product constant. This $K_{sp}$ expression for the silver chloride example can be written as follows.

$$K_{sp} = [Ag^+][Cl^-]$$

Note that the coefficients in the balanced equation are understood to be 1 for both silver and chlorine. Therefore, no exponents appear in the $K_{sp}$ expression. As with any equilibrium expression, the concentration of each component is raised to the power of its coefficient from the balanced chemical equation. The value for this solubility product constant is $1.77 \times 10^{-10}$ at 25°C. The very small value of $K_{sp}$ indicates that silver chloride is only very slightly soluble in aqueous solution at this temperature. The value of $K_{sp}$ supports the observation that little seems to occur when silver chloride is stirred into a water solution.

In this worksheet, you will learn to apply the solubility equilibrium relationship to determine $K_{sp}$ for substances and to calculate concentrations of ions in saturated solutions using their $K_{sp}$ values.

### General Plan for Solving Solubility Equilibrium Problems

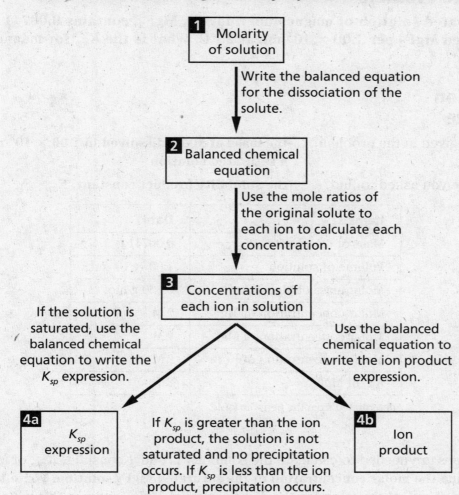

**1** Molarity of solution

Write the balanced equation for the dissociation of the solute.

**2** Balanced chemical equation

Use the mole ratios of the original solute to each ion to calculate each concentration.

**3** Concentrations of each ion in solution

If the solution is saturated, use the balanced chemical equation to write the $K_{sp}$ expression.

Use the balanced chemical equation to write the ion product expression.

**4a** $K_{sp}$ expression

If $K_{sp}$ is greater than the ion product, the solution is not saturated and no precipitation occurs. If $K_{sp}$ is less than the ion product, precipitation occurs.

**4b** Ion product

# Sample Problem 1

A saturated solution of magnesium fluoride, $MgF_2$, contains 0.00741 g of dissolved $MgF_2$ per $1.00 \times 10^2$ mL at 25°C. What is the $K_{sp}$ for magnesium fluoride?

# Solution
## ANALYZE

What is given in the problem?     the mass of $MgF_2$ dissolved in $1.00 \times 10^2$ mL of a saturated solution

What are you asked to find?     the solubility product constant, $K_{sp}$

| Items | Data |
|---|---|
| Mass of dissolved $MgF_2$ | 0.00741 g |
| Volume of solution | $1.00 \times 10^2$ mL |
| Molar mass of $MgF_2$* | 62.30 g/mol |
| Molar concentration of $MgF_2$ | ? M |
| Molar concentration of $Mg^{2+}$ | ? M |
| Molar concentration of $F^-$ | ? M |
| $K_{sp}$ of $MgF_2$ | ? |

*determined from the periodic table

## PLAN

What steps are needed to calculate the solubility product constant, $K_{sp}$, of $MgF_2$? Determine the molar concentration of the saturated $MgF_2$ solution. Write the balanced chemical equation for the dissociation of $MgF_2$, and use this equation to determine the concentrations of each ion in solution. Compute $K_{sp}$.

**Problem Solving** *continued*

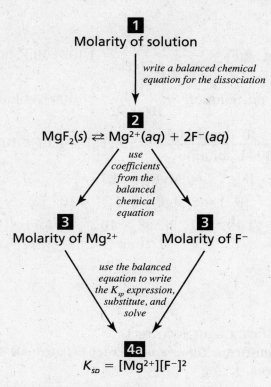

**1**

**Molarity of solution**

*write a balanced chemical
equation for the dissociation*

**2**

$MgF_2(s) \rightleftarrows Mg^{2+}(aq) + 2F^-(aq)$

*use
coefficients
from the
balanced
chemical
equation*

**3** | **3**
**Molarity of $Mg^{2+}$** | **Molarity of $F^-$**

*use the balanced
equation to write
the $K_{sp}$ expression,
substitute, and
solve*

**4a**

$K_{sp} = [Mg^{2+}][F^-]^2$

Calculate the molarity of the saturated $MgF_2$ solution.

$$\text{g } \overset{given}{MgF_2} \times \frac{\overset{\frac{1}{molar\ mass\ MgF_2}}{1 \text{ mol } MgF_2}}{62.30 \text{ g } MgF_2} = \text{mol } MgF_2$$

$$\overset{given}{\text{mL solution}} \times \frac{1 \text{ L}}{1000 \text{ mL}} = \text{L solution}$$

$$\frac{\overset{calculated\ above}{\text{mol } MgF_2}}{\underset{calculated\ above}{\text{L solution}}} = [MgF_2]$$

Write the balanced chemical equation for the dissociation to determine the mole ratios of solute and ions.

$$MgF_2(s) \overset{a}{} Mg^{2+}(aq) + 2F^-(aq)$$
$$[MgF_2] = [Mg^{2+}]$$
$$2[MgF_2] = [F^-]$$

Write the $K_{sp}$ expression.

$$K_{sp} = [Mg^{2+}][F^-]^2$$
$$K_{sp} = [MgF_2]\,\overset{calculated\ above}{(2[MgF_2])^2}$$

**▌ Problem Solving** *continued*

## COMPUTE

$$0.007\,41 \text{ g MgF}_2 \times \frac{1 \text{ mol MgF}_2}{62.30 \text{ g MgF}_2} = 1.19 \times 10^{-4} \text{ mol MgF}_2$$

$$100. \text{ mL solution} \times \frac{1 \text{ L}}{1000 \text{ mL}} = 0.100 \text{ L solution}$$

$$\frac{1.19 \times 10^{-4} \text{ mol MgF}_2}{0.100 \text{ L solution}} = [\text{MgF}_2] = 1.19 \times 10^{-3} \text{ M}$$

$$[\text{Mg}^{2+}] = [\text{MgF}_2] = 1.19 \times 10^{-3} \text{ M}$$

$$[\text{F}^-] = 2[\text{MgF}_2] = 2.38 \times 10^{-3} \text{ M}$$

$$K_{sp} = [1.19 \times 10^{-3}][2.38 \times 10^{-3}]^2 = 6.74 \times 10^{-9}$$

## EVALUATE

Are the units correct?

**Yes; $K_{sp}$ has no units.**

Is the number of significant figures correct?

**Yes; the number of significant figures is correct because all data were given to three significant figures.**

Is the answer reasonable?

**Yes; the calculation can be approximated as $(1 \times 10^{-3})(2.5 \times 10^{-3})^2 = 6 \times 10^{-9}$, which is of the same order of magnitude as the calculated answer.**

## Practice

**1.** Silver bromate, $AgBrO_3$, is slightly soluble in water. A saturated solution is found to contain 0.276 g $AgBrO_3$ dissolved in 150.0 mL of water. Calculate $K_{sp}$ for silver bromate. ans: $K_{sp} = 6.09 \times 10^{-5}$

**2.** 2.50 L of a saturated solution of calcium fluoride leaves a residue of 0.0427 g of $CaF_2$ when evaporated to dryness. Calculate the $K_{sp}$ of $CaF_2$. ans: $K_{sp} = 4.20 \times 10^{-11}$

**Problem Solving** *continued*

# Sample Problem 2

The $K_{sp}$ for lead(II) iodide is $7.08 \times 10^{-9}$ at 25°C. What is the molar concentration of $PbI_2$ in a saturated solution?

# Solution

## ANALYZE

What is given in the problem    **the solubility product constant, $K_{sp}$ of $PbI_2$**

What are you asked to find?    **the concentration of $PbI_2$ in a saturated solution**

| Items | Data |
|---|---|
| Ksp of $PbI_2$ | $7.08 \times 10^{-9}$ |
| Concentration of $Pb^{2+}$ | ? |
| Concentration of $I^-$ | ? |
| Concentration of $PbI_2$ in solution | ? |

## PLAN

What steps are needed to calculate the concentration of dissolved $PbI_2$ in a saturated solution?

**Write the equation for the dissociation of $PbI_2$. Set up the equation for $K_{sp}$, and compute the concentrations of the ions. Determine the concentration of dissolved solute.**

Write the balanced chemical equation for the dissociation of lead(II) iodide, $PbI_2$ in aqueous solution.

$$PbI_2(s) \rightleftharpoons Pb^{2+}(aq) + 2I^-(aq)$$

Write the $K_{sp}$ expression.

$$K_{sp} = [Pb^{2+}][I^-]^2$$

Substitute $x$ for $[Pb^{2+}]$. The balanced equation gives the following relationship:

$$2[Pb^{2+}] = [I^-].$$

Therefore, $[I^-] = 2x$.

$$K_{sp} = [x][2x]^2$$

Rearrange, and solve for $x$.

$$K_{sp} = [x][4x^2]$$
$$K_{sp} = 4x^3$$

$$x = \sqrt[3]{\frac{\overset{given}{K_{sp}}}{4}} = [Pb^{2+}]$$

Relate the substituted value to the unknown solution concentration using the mole ratio from the original balanced chemical equation. The mole ratio shows that $[Pb^{2+}] = [PbI_2]$.

**COMPUTE**

$$x = \sqrt[3]{\frac{7.08 \times 10^{-9}}{4}} = [Pb^{2+}] = 1.21 \times 10^{-3} \text{ M}$$

$$[Pb^{2+}] = [PbI_2] = 1.21 \times 10^{-3} \text{ M}$$

**EVALUATE**

Are the units correct?
**Yes; concentrations are in molarity (mol/L).**

Is the number of significant figures correct?
**Yes; the number of significant figures is correct because data were given to three significant figures.**

Is the answer reasonable?
**Yes; the best check is to use the result to calculate $K_{sp}$ and see if it gives (or is very near) the $K_{sp}$ you started with. In this case, the calculated $K_{sp}$ is $7.08 \times 10^{-9}$, the same value as was given.**

## Practice

1. The $K_{sp}$ of calcium sulfate, $CaSO_4$, is $9.1 \times 10^{-6}$. What is the molar concentration of $CaSO_4$ in a saturated solution? **ans: $3.0 \times 10^{-3}$ M**

2. A salt has the formula $X_2Y$, and its $K_{sp}$ is $4.25 \times 10^{-7}$.

   **a.** What is the molarity of a saturated solution of the salt?
   **ans: $[X_2Y] = 4.74 \times 10^{-3}$ M**

   **b.** What is the molarity of a solution of AZ if its $K_{sp}$ is the same value?
   **ans: $[AZ] = 6.52 \times 10^{-4}$ M**

# Sample Problem 3

Will precipitation of strontium sulfate occur when 50.0 mL of 0.025 M strontium nitrate solution is mixed with 50.0 mL of a 0.014 M copper(II) sulfate solution? The $K_{sp}$ of strontium nitrate is $3.2 \times 10^{-7}$.

# Solution

## ANALYZE

What is given in the problem?  **the molar concentrations of the solutions to be mixed, the identities of the solutes, and the volumes of the solutions to be mixed**

What are you asked to find?  **whether a precipitate of strontium sulfate forms when the two solutions are mixed**

| Items | Data |
|---|---|
| Concentration of solution 1 | 0.025 M SrNO$_3$ |
| Volume of solution 1 | 50.0 mL |
| Concentration of solution 2 | 0.014 M CuSO$_4$ |
| Volume of solution 2 | 50.0 mL |
| Volume of combined solution | 100.0 mL |
| Concentration of combined solution | ? M SrSO$_4$ |
| Potential precipitate | SrSO$_4$ |
| $K_{sp}$ of SrSO$_4$ | $3.2 \times 10^{-7}$ |
| Precipitate/no precipitate forms | ? |

## PLAN

What steps are needed to determine whether a precipitate will form?
**Calculate the molar concentrations of the ions that can form a precipitate in the new volume of solution. Use these concentrations to calculate the ion product. Compare the ion product with $K_{sp}$.**

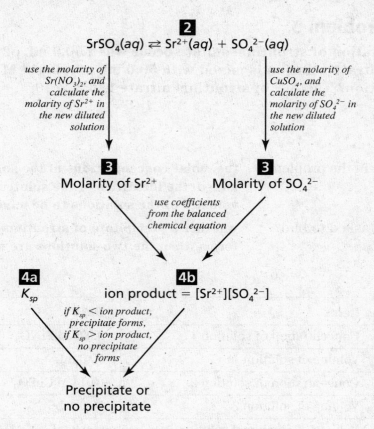

**2**
$$SrSO_4(aq) \rightleftharpoons Sr^{2+}(aq) + SO_4^{2-}(aq)$$

*use the molarity of Sr(NO₃)₂, and calculate the molarity of Sr²⁺ in the new diluted solution*

*use the molarity of CuSO₄, and calculate the molarity of SO₄²⁻ in the new diluted solution*

**3** Molarity of $Sr^{2+}$     **3** Molarity of $SO_4^{2-}$

*use coefficients from the balanced chemical equation*

**4a** $K_{sp}$     **4b** ion product $= [Sr^{2+}][SO_4^{2-}]$

*if $K_{sp} <$ ion product, precipitate forms, if $K_{sp} >$ ion product, no precipitate forms*

**Precipitate or no precipitate**

Write the balanced equation for the dissociation of $SrSO_4$.

$$SrSO_4(aq) \rightleftharpoons Sr^{2+}(aq) + SO_4^{2-}(aq)$$

Calculate the molarities of $Sr^{2+}$ and $SO_4^{2-}$. This is a simple dilution calculation. The subscript 1 in each case represents that individual solution; the subscript 2 represents the combined solution.

$$Sr(NO_3)_2(aq) \rightleftharpoons Sr^{2+}(aq) + 2NO_3^-(aq)$$
$$[Sr^{2+}]_1 = [Sr(NO_3)_2]$$
$$[Sr^{2+}]_1 V_1 = [Sr^{2+}]_2 V_2$$

$$[Sr^{2+}]_2 = \frac{\overset{\text{calculated above}}{[Sr^{2+}]_1} \overset{\text{given}}{V_1}}{\underset{\substack{\text{sum of volumes of} \\ \text{solutions mixed}}}{V_2}}$$

$$CuSO_4(aq) \rightleftharpoons Cu^{2+}(aq) + SO_4^{2-}(aq)$$
$$[SO_4^{2-}]_1 = [CuSO_4]$$
$$[SO_4^{2-}]_1 V_1 = [SO_4^{2-}]_2 V_2$$

$$[SO_4^{2-}]_2 = \frac{\overset{\text{calculated above}}{[SO_4^{2-}]_1} \overset{\text{given}}{V_1}}{\underset{\substack{\text{sum of volumes of} \\ \text{solutions mixed}}}{V_2}}$$

Calculate the ion product for $SrSO_4$.

$$\text{ion product} = [Sr^{2+}]_2^{\overset{calculated}{\overset{above}{}}}[SO_4^{2-}]_2$$

Compare the ion product to the $K_{sp}$ value to determine if precipitation occurs.

**COMPUTE**

$$[Sr^{2+}]_2 = \frac{0.025 \text{ M} \times 50.0 \text{ mL}}{100.0 \text{ mL}} = 1.2 \times 10^{-2} \text{ M}$$

$$[SO_4^{2-}]_2 = \frac{0.014 \text{ M} \times 50.0 \text{ mL}}{100.0 \text{ mL}} = 7.0 \times 10^{-3} \text{ M}$$

$$\text{ion product} = [1.2 \times 10^{-2}][7.0 \times 10^{-3}] = 8.4 \times 10^{-5}$$
$$K_{sp} = 3.2 \times 10^{-7}$$
$$K_{sp} < \text{ion product}$$

Precipitation will occur.

**EVALUATE**

Are the units correct?
**Yes; the ion product has no units.**

Is the number of significant figures correct?
**Yes; the number of significant figures is correct because data were given to a minimum of two significant figures.**

Is the answer reasonable?
**Yes; the calculation can be approximated as $0.01 \times 0.007 = 0.000\ 07 = 7 \times 10^{-5}$, which is of the same order of magnitude as the calculated result.**

# Practice

**In each of the following problems, include the calculated ion product with your answer.**

1. Will a precipitate of $Ca(OH)_2$ form when 320. mL of a 0.046 M solution of NaOH mixes with 400. mL of a 0.085 M $CaCl_2$ solution? $K_{sp}$ of $Ca(OH)_2$ is $5.5 \times 10^{-6}$. **ans: ion product = $1.9 \times 10^{-5}$, precipitation occurs.**

**▌Problem Solving** *continued*

**2.** 20.00 mL of a 0.077 M solution of silver nitrate, $AgNO_3$, is mixed with 30.00 mL of a 0.043 M solution of sodium acetate, $NaC_2H_3O_2$. Does a precipitate form? The $K_{sp}$ of $AgC_2H_3O_2$ is $2.5 \times 10^{-3}$.
**ans: ion product = $8.1 \times 10^{-4}$, no precipitate**

**3.** If you mix 100. mL of 0.036 M $Pb(C_2H_3O_2)_2$ with 50. mL of 0.074 M NaCl, will a precipitate of $PbCl_2$ form? $K_{sp}$ of $PbCl_2$ is $1.9 \times 10^{-4}$.
**ans: ion product = $1.5 \times 10^{-5}$, no precipitate**

**4.** If 20.00 mL of a 0.0090 M solution of $(NH_4)_2S$ is mixed with 120.00 mL of a 0.0082 M solution of $Al(NO_3)_3$, does a precipitate form? The $K_{sp}$ of $Al_2S_3$ is $2.00 \times 10^{-7}$. **ans: ion product = $1.1 \times 10^{-13}$, no precipitate**

| **Problem Solving** *continued* |

# Additional Problems—Equilibrium of Salts, $K_{sp}$

1. The molar concentration of a saturated calcium chromate, $CaCrO_4$, solution is 0.010 M at 25°C. What is the $K_{sp}$ of calcium chromate?

2. A 10.00 mL sample of a saturated lead selenate solution is found to contain 0.00136 g of dissolved $PbSeO_4$ at 25°C. Determine the $K_{sp}$ of lead selenate.

3. A 22.50 mL sample of a saturated copper(I) thiocyanate, CuSCN, solution at 25°C is found to have a $4.0 \times 10^{-6}$ M concentration.

   **a.** Determine the $K_{sp}$ of CuSCN.

   **b.** What mass of CuSCN would be dissolved in $1.0 \times 10^3$ L of solution?

4. A saturated solution of silver dichromate, $Ag_2Cr_2O_7$, has a concentration of $3.684 \times 10^{-3}$ M. Calculate the $K_{sp}$ of silver dichromate.

5. The $K_{sp}$ of barium sulfite, $BaSO_3$, at 25°C is $8.0 \times 10^{-7}$.

   **a.** What is the molar concentration of a saturated solution of $BaSO_3$?

   **b.** What mass of $BaSO_3$ would dissolve in 500. mL of water?

6. The $K_{sp}$ of lead(II) chloride at 25°C is $1.9 \times 10^{-4}$. What is the molar concentration of a saturated solution at 25°C?

7. The $K_{sp}$ of barium carbonate at 25°C is $1.2 \times 10^{-8}$.

   **a.** What is the molar concentration of a saturated solution of $BaCO_3$ at 25°C?

   **b.** What volume of water would be needed to dissolve 0.10 g of barium carbonate?

8. The $K_{sp}$ of $SrSO_4$ is $3.2 \times 10^{-7}$ at 25°C.

   **a.** What is the molar concentration of a saturated $SrSO_4$ solution?

   **b.** If 20.0 L of a saturated solution of $SrSO_4$ were evaporated to dryness, what mass of $SrSO_4$ would remain?

9. The $K_{sp}$ of strontium sulfite, $SrSO_3$, is $4.0 \times 10^{-8}$ at 25°C. If 1.0000 g of $SrSO_3$ is stirred in 5.0 L of water until the solution is saturated and then filtered, what mass of $SrSO_3$ would remain?

10. The $K_{sp}$ of manganese(II) arsenate is $1.9 \times 10^{-11}$ at 25°C. What is the molar concentration of $Mn_3(AsO_4)_2$ in a saturated solution? Note that five ions are produced from the dissociation of $Mn_3(AsO_4)_2$.

11. Suppose that 30.0 mL of a 0.0050 M solution of $Sr(NO_3)_2$ is mixed with 20.0 mL of a 0.010 M solution of $K_2SO_4$ at 25°C. The $K_{sp}$ of $SrSO_4$ is $3.2 \times 10^{-7}$.

   **a.** What is the ion product of the ions that can potentially form a precipitate?

   **b.** Does a precipitate form?

12. Lead(II) bromide, $PbBr_2$, is slightly soluble in water. Its $K_{sp}$ is $6.3 \times 10^{-6}$ at 25°C. Suppose that 120. mL of a 0.0035 M solution of $MgBr_2$ is mixed with 180. mL of a 0.0024 M $Pb(C_2H_3O_2)_2$ solution at 25°C.

   **a.** What is the ion product of $Br^-$ and $Pb^{2+}$ in the mixed solution?

   **b.** Does a precipitate form?

**| Problem Solving** *continued*

13. The $K_{sp}$ of $Mg(OH)_2$ at 25°C is $1.5 \times 10^{-11}$.

   **a.** Write the equilibrium equation for the dissociation of $Mg(OH)_2$.

   **b.** What volume of water would be required to dissolve 0.10 g of $Mg(OH)_2$?

   **c.** Considering that magnesium hydroxide is essentially insoluble, why is it possible to titrate a suspension of $Mg(OH)_2$ to an equivalence point with a strong acid such as HCl?

14. Lithium carbonate is somewhat soluble in water; its $K_{sp}$ at 25°C is $2.51 \times 10^{-2}$.

   **a.** What is the molar concentration of a saturated $Li_2CO_3$ solution?

   **b.** What mass of $Li_2CO_3$ would you dissolve in order to make 3440 mL of saturated solution?

15. A 50.00 mL sample of a saturated solution of barium hydroxide, $Ba(OH)_2$, is titrated to the equivalence point by 31.61 mL of a 0.3417 M solution of HCl. Determine the $K_{sp}$ of $Ba(OH)_2$.

16. Calculate the $K_{sp}$ for salts represented by QR that dissociate into two ions, $Q^+$ and $R^-$, in each of the following solutions:

   **a.** saturated solution of QR is 1.0 M

   **b.** saturated solution of QR is 0.50 M

   **c.** saturated solution of QR is 0.1 M

   **d.** saturated solution of QR is 0.001 M

17. Suppose that salts QR, $X_2Y$, $KL_2$, $A_3Z$, and $D_2E_3$ form saturated solutions that are 0.02 M in concentration. Calculate $K_{sp}$ for each of these salts.

18. The $K_{sp}$ at 25°C of silver bromide is $5.0 \times 10^{-13}$. What is the molar concentration of a saturated AgBr solution? What mass of silver bromide would dissolve in 10.0 L of saturated solution at 25°C?

19. The $K_{sp}$ at 25°C for calcium hydroxide is $5.5 \times 10^{-6}$.

   **a.** Calculate the molarity of a saturated $Ca(OH)_2$ solution.

   **b.** What is the $OH^-$ concentration of this solution?

20. The $K_{sp}$ of magnesium carbonate is $3.5 \times 10^{-8}$ at 25°C. What mass of $MgCO_3$ would dissolve in 4.00 L of water at 25°C?

Skills Worksheet

# Problem Solving

## pH

In 1909, Danish biochemist S. P. L Sørensen introduced a system in which acidity was expressed as the negative logarithm of the $H^+$ concentration. In this way, the acidity of a solution having $H^+$ concentration of $10^{-5}$ M would have a value of 5. Because the *power* of 10 was now a part of the number, the system was called *pH*, meaning *power of hydrogen.*

Taking the negative logarithm of the hydronium ion concentration will give you a solution's pH as given by the following equation.

$$pH = -\log[H_3O^+]$$

Likewise, the value pOH equals the negative logarithm of the hydroxide ion concentration.

$$pOH = -\log[OH^-]$$

Water molecules interact with each other and ionize. At the same time, the ions in solution reform molecules of water. This process is represented by the following reversible equation.

$$H_2O(l) + H_2O(l) \rightleftarrows H_3O(aq) + OH^-(aq)$$

In pure water the concentrations of hydroxide ions and hydronium ions will always be equal. These two quantities are related by a term called the *ion product constant* for water, $K_w$.

$$K_w = [H_3O^+][OH^-]$$

The ion product constant for water can be used to convert from pOH to pH. The following equations derive a simple formula for this conversion.

Pure water has a pH of 7. Rearranging the equation for pH, you can solve for the hydronium ion concentration of pure water, which is equal to the hydroxide ion concentration. These values can be used to obtain a numerical value for $K_w$.

$$K_w = [H_3O^+][OH^-] = [1 \times 10^{-7}][1 \times 10^{-7}] = 1 \times 10^{-14}$$

Rearrange the $K_w$ expression to solve for $[OH^-]$.

$$[OH^-] = \frac{K_w}{[H_3O^+]}$$

Substitute this value into the equation for pOH.

$$pOH = -\log\frac{K_w}{[H_3O^+]}$$

The logarithm of a quotient is the difference of the logarithms of the numerator and the denominator.

$$pOH = -\log K_w + \log[H_3O^+]$$

Substitute the value for $K_w$ and pH for $+\log[H_3O^+]$.

$$pOH = -\log(1.0 \times 10^{-14}) - pH$$

**Problem Solving** *continued*

The negative logarithm of $10^{-14}$ is 14. Substitute this value, and rearrange.

$$\text{pOH} + \text{pH} = 14$$

### General Plan for Solving pH Problems

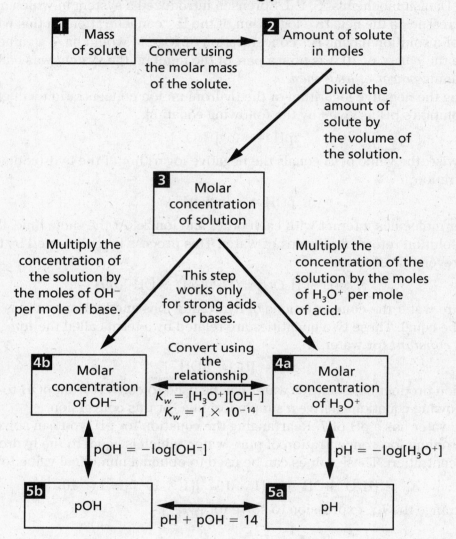

# Sample Problem 1

A HCl solution has a concentration of 0.0050 M. Calculate $[OH^-]$ and $[H_3O^+]$ for this solution. HCl is a strong acid, so assume it is 100% ionized.

# Solution

## ANALYZE

What is given in the problem?     **the molarity of the HCl solution, and the fact that HCl is a strong acid**

What are you asked to find?     $[H_3O^+]$ **and** $[OH^-]$

| Items | Data |
|---|---|
| Identity of solute | HCl |
| Concentration of solute | 0.0050 M |
| Acid or base | acid |
| $K_w$ | $1.0 \times 10^{-14}$ |
| $[H_3O^+]$ | ? M |
| $[OH^-]$ | ? M |

## PLAN

What steps are needed to calculate the concentration of $H_3O^+$ and $OH^-$?
**Determine $[H_3O^+]$ from molarity and the fact that the acid is strong. Use $K_w$ to calculate $[OH^-]$.**

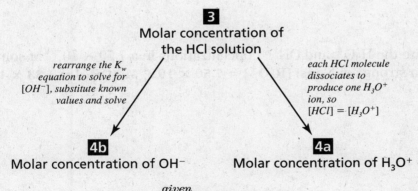

**3**

Molar concentration of the HCl solution

*rearrange the $K_w$ equation to solve for $[OH^-]$, substitute known values and solve*

*each HCl molecule dissociates to produce one $H_3O^+$ ion, so $[HCl] = [H_3O^+]$*

**4b**
Molar concentration of $OH^-$

**4a**
Molar concentration of $H_3O^+$

$$\overset{given}{[HCl]} = [H_3O^+]$$

$$1 \times \overset{K_w}{10^{-14}} = [OH^-] \times \overset{\substack{calculated \\ above}}{[H_3O^+]}$$

$$[OH^-] = \frac{1 \times \overset{K_w}{10^{-14}}}{\underset{calculated\ above}{[H_3O^+]}}$$

**Problem Solving** *continued*

## COMPUTE

$$[H_3O^+] = 0.0050 \text{ M}$$

$$[OH^-] = \frac{1.0 \times 10^{-14}}{0.0050} = 2.0 \times 10^{-12} \text{ M}$$

## EVALUATE

Are the units correct?

**Yes; molarity, or mol/L, was required.**

Is the number of significant figures correct?

**Yes; the number of significant figures is correct because molarity of HCl was given to two significant figures.**

Is the answer reasonable?

**Yes; the two concentrations multiply to give $10 \times 10^{-15}$, which is equal to $1.0 \times 10^{-14}$.**

# Practice

1. The hydroxide ion concentration of an aqueous solution is $6.4 \times 10^{-5}$ M. What is the hydronium ion concentration? **ans: $1.6 \times 10^{-10}$ M**

2. Calculate the $H_3O^+$ and $OH^-$ concentrations in a $7.50 \times 10^{-4}$ M solution of $HNO_3$, a strong acid. **ans: $[H_3O^+] = 7.50 \times 10^{-4}$ M; $[OH^-] = 1.33 \times 10^{-11}$ M**

**Problem Solving** *continued*

# Sample Problem 2

**Calculate the pH of a 0.000 287 M solution of $H_2SO_4$. Assume 100% ionization.**

# Solution

## ANALYZE

What is given in the problem?    **the molarity of the $H_2SO_4$ solution, and the fact that $H_2SO_4$ is completely ionized**

What are you asked to find?    **pH**

| Items | Data |
|---|---|
| Identity of solute | $H_2SO_4$ |
| Concentration of solute | 0.000 287 M |
| Acid or base | acid |
| $K_w$ | $1.0 \times 10^{-14}$ |
| $[H_3O^+]$ | ? M |
| pH | ? |

## PLAN

What steps are needed to calculate the concentration of $H_3O^+$ and the pH?
**Determine $[H_3O^+]$ from molarity and the fact that the acid is 100% ionized. Determine the pH, negative logarithm of the concentration.**

**3**
Molar concentration of
the $H_2SO_4$ solution

*each $H_2SO_4$ molecule dissociates to produce two $H_3O^+$ ions, so $2 \times [H_2SO_4] = [H_3O^+]$*

**4a**
Molar concentration of $H_3O^+$

*$pH = -log[H_3O^+]$*

**5a**
pH

*given*
$$2 \times [H_2SO_4] = [H_3O^+]$$

*calculated above*
$$-log[H_3O^+] = pH$$

## COMPUTE

$$[H_3O^+] = 2 \times 0.000\ 287\ M = 5.74 \times 10^{-4}\ M$$
$$-\log[5.74 \times 10^{-4}] = 3.24$$

## EVALUATE

Are the units correct?

**Yes; there are no units on a pH value.**

Is the number of significant figures correct?

**Yes; the number of significant figures is correct because molarity of $H_2SO_4$ was given to three significant figures.**

Is the answer reasonable?

**Yes. You would expect the pH of a dilute $H_2SO_4$ solution to be below 7.**

# Practice

**1.** Determine the pH of a 0.001 18 M solution of HBr. **ans: 2.93**

**2.** What is the pH of a solution that has a hydronium ion concentration of 1.0 M? **ans: 0.00**

**Problem Solving** *continued*

**3.** What is the pH of a 2.0 M solution of HCl, assuming the acid remains 100% ionized? **ans: −0.30**

**4.** What is the theoretical pH of a 10.0 M solution of HCl? **ans: −1.00**

## Sample Problem 3

A solution of acetic acid has a pH of 5.86. What are the pOH and [OH⁻] of the solution?

## Solution

### ANALYZE

What is given in the problem?     **the pH of the acetic acid solution**

What are you asked to find?     **pOH and [OH⁻]**

| Items | Data |
|---|---|
| Identity of solute | acetic acid |
| Acid or base | acid |
| pH | 5.86 |
| pOH | ? |
| [OH⁻] | ? M |

### PLAN

What steps are needed to calculate the pOH?
**The sum of the pH and pOH of any solution is 14.00. Use this relationship to find the pOH of the acetic acid solution.**

What steps are needed to calculate [OH⁻]?
**The pOH of a solution is the negative logarithm of the hydroxide ion concentration. Therefore, calculate [OH⁻] using the inverse logarithm of the negative pOH.**

**4b**
Molar concentration of [OH⁻]

*convert using the relationship* $pOH = -\log[OH^-]$

**5b**                                    **5a**
pOH ← *convert using the relationship* pH
$pH + pOH = 14.00$

*given*
$$pH + pOH = 14.00$$

*given*
$$14.00 - pH = pOH$$

*calculated above*
$$pOH = -\log[OH^-]$$

*calculated above*
$$10^{-pOH} = [OH^-]$$

## COMPUTE

$$14.00 - 5.86 = 8.14$$
$$10^{-8.14} = 7.2 \times 10^{-9} \text{ M}$$

## EVALUATE

Are the units correct?

**Yes; there are no units on a pOH value, and [OH⁻] has the correct units of molarity.**

Is the number of significant figures correct?

**Yes; the number of significant figures is correct because the data were given to three significant figures.**

Is the answer reasonable?

**Yes; you would expect the pOH of an acid to be above 7, and the hydroxide ion concentration to be small.**

# Practice

**1.** What is the pH of a solution with the following hydroxide ion concentrations?

   **a.** $1 \times 10^{-5}$ M **ans: 9.0**

   **b.** $5 \times 10^{-8}$ M **ans: 6.7**

   **c.** $2.90 \times 10^{-11}$ M **ans: 3.46**

**2.** What are the pOH and hydroxide ion concentration of a solution with a pH of 8.92? **ans: pOH = 5.08, [OH⁻] = 8.3 × 10⁻⁶ M**

**3.** What are the pOH values of solutions with the following hydronium ion concentrations?

   **a.** $2.51 \times 10^{-13}$ M **ans: 1.40**

   **b.** $4.3 \times 10^{-3}$ M **ans: 11.6**

   **c.** $9.1 \times 10^{-6}$ M **ans: 8.96**

   **d.** 0.070 M **ans: 12.8**

## Sample Problem 4

Determine the pH of a solution made by dissolving 4.50 g NaOH in a
0.400 L aqueous solution. NaOH is a strong base.

## Solution

### ANALYZE

What is given in the problem?    **the mass of NaOH, and the solution volume**

What are you asked to find?    **pH**

| Items | Data |
|-------|------|
| Identity of solute | NaOH |
| Mass of solute | 4.50 g |
| Molar mass of solute | 40.00 g/mol |
| Volume of solution | 0.400 L |
| Concentration of solute | ? M |
| Acid or base | base |
| $[OH^-]$ | ? M |
| pOH | ? |
| pH | ? |

### PLAN

What steps are needed to calculate the pH?
**First determine the concentration of the solution. Then find the concentration
of hydroxide ions. Calculate the pOH of the solution, and use this to find the pH.**

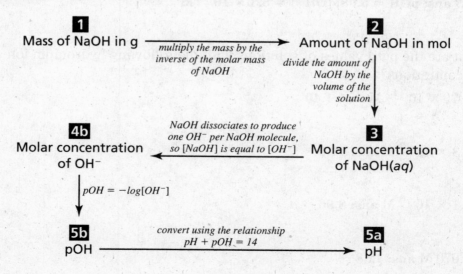

**| Problem Solving** *continued*

$$\text{g } \overset{given}{\text{NaOH}} \times \frac{\overset{\frac{1}{\text{molar mass of NaOH}}}{1 \text{ mol NaOH}}}{40.00 \text{ g NaOH}} \times \frac{1}{\underset{given}{\text{L solution}}} \times \frac{1 \text{ mol OH}^-}{1 \text{ mol NaOH}} = [\text{OH}^-]$$

$$\text{pOH} = -\log[\overset{\underset{above}{calculated}}{\text{OH}^-}]$$

$$\text{pH} + \overset{\underset{above}{calculated}}{\text{pOH}} = 14.00$$

$$14.00 - \overset{\underset{above}{calculated}}{\text{pOH}} = \text{pH}$$

## COMPUTE

$$4.50 \text{ g NaOH} \times \frac{1 \text{ mol NaOH}}{40.00 \text{ g NaOH}} \times \frac{1}{0.400 \text{ L solution}} \times \frac{1 \text{ mol OH}^-}{1 \text{ mol NaOH}} = 0.281 \text{ M}$$

$$-\log[0.281] = 0.551$$

$$14.00 - 0.551 = 13.45$$

## EVALUATE

Are the units correct?
**Yes; pH has no units.**

Is the number of significant figures correct?
**Yes; the number of significant figures is correct because the value 14.00 has two decimal places.**

Is the answer reasonable?
**Yes; NaOH is a strong base, so you would expect it to have a pH around 14.**

# Practice

**1.** A solution is prepared by dissolving 3.50 g of sodium hydroxide in water and adding water until the total volume of the solution is 2.50 L. What are the $\text{OH}^-$ and $\text{H}_3\text{O}^+$ concentrations? **ans: [OH$^-$] = 0.0350 M, [H$_3$O$^+$] = 2.86 × 10$^{-13}$ M**

**2.** If 1.00 L of a potassium hydroxide solution with a pH of 12.90 is diluted to 2.00 L, what is the pH of the resulting solution? **ans: 13.20**

# Additional Problems—pH

1. Calculate the $H_3O^+$ and $OH^-$ concentrations in the following solutions. Each is either a strong acid or a stong base.

   **a.** 0.05 M sodium hydroxide

   **b.** 0.0025 M sulfuric acid

   **c.** 0.013 M lithium hydroxide

   **d.** 0.150 M nitric acid

   **e.** 0.0200 M calcium hydroxide

   **f.** 0.390 M perchloric acid

   **g.** What is the pH of each solution in items 1a. to 1f.?

2. Calculate $[H_3O^+]$ and $[OH^-]$ in a 0.160 M solution of potassium hydroxide. Assume that the solute is 100% dissociated at this concentration.

3. The pH of an aqueous solution of sodium hydroxide is 12.9. What is the molarity of the solution?

4. What is the pH of a 0.001 25 M HBr solution? If 175 mL of this solution is diluted to a total volume of 3.00 L, what is the pH of the diluted solution?

5. What is the pH of a 0.0001 M solution of NaOH? What is the pH of a 0.0005 M solution of NaOH?

6. A solution is prepared using 15.0 mL of 1.0 M HCl and 20.0 mL of 0.50 M $HNO_3$. The final volume of the solution is 1.25 L. Answer the following questions.

   **a.** What are the $[H_3O^+]$ and $[OH^-]$ in the final solution?

   **b.** What is the pH of the final solution?

7. A container is labeled 500.0 mL of 0.001 57 M nitric acid solution. A chemist finds that the container was not sealed and that some evaporation has taken place. The volume of solution is now 447.0 mL.

   **a.** What was the original pH of the solution?

   **b.** What is the pH of the solution now?

8. Calculate the hydroxide ion concentration in an aqueous solution that has a 0.000 35 M hydronium ion concentration.

9. A solution of sodium hydroxide has a pH of 12.14. If 50.00 mL of the solution is diluted to 2.000 L with water, what is the pH of the diluted solution?

10. An acetic acid solution has a pH of 4.0. What are the $[H_3O^+]$ and $[OH^-]$ in this solution?

11. What is the pH of a 0.000 460 M solution of $Ca(OH)_2$?

12. A solution of strontium hydroxide with a pH of 11.4 is to be prepared. What mass of $Sr(OH)_2$ would be required to make 1.00 L of this solution?

13. A solution of $NH_3$ has a pH of 11.00. What are the concentrations of hydronium and hydroxide ions in this solution?

**14.** Acetic acid does not completely ionize in solution. Percent ionization of a substance dissolved in water is equal to the moles of ions produced as a percentage of the moles of ions that would be produced if the substance were completely ionized. Calculate the percent ionization of acetic acid the following solutions.

**a.** 1.0 M acetic acid solution with a pH of 2.40

**b.** 0.10 M acetic acid solution with a pH of 2.90

**c.** 0.010 M acetic acid solution, with a pH of 3.40

**15.** Calculate the pH of an aqueous solution that contains 5.00 g of $HNO_3$ in 2.00 L of solution.

**16.** A solution of HCl has a pH of 1.50. Determine the pH of the solutions made in each of the following ways.

**a.** 1.00 mL of the solution is diluted to 1000. mL with water.

**b.** 25.00 mL is diluted to 200 mL with distilled water.

**c.** 18.83 mL of the solution is diluted to 4.000 L with distilled water.

**d.** 1.50 L is diluted to 20.0 kL with distilled water.

**17.** An aqueous solution contains 10 000 times more hydronium ions than hydroxide ions. What is the concentration of each ion?

**18.** A potassium hydroxide solution has a pH of 12.90. Enough acid is added to react with half of the $OH^-$ ions present. What is the pH of the resulting solution? Assume that the products of the neutralization have no effect on pH and that the amount of additional water produced is negligible.

**19.** A hydrochloric acid solution has a pH of 1.70. What is the $[H_3O^+]$ in this solution? Considering that HCl is a strong acid, what is the HCl concentration of the solution?

**20.** What is the molarity of a solution of the strong base $Ca(OH)_2$ in a solution that has a pH of 10.80?

**21.** You have a 1.00 M solution of the strong acid, HCl. What is the pH of this solution? You need a solution of pH 4.00. To what volume would you dilute 1.00 L of the HCl solution to get this pH? To what volume would you dilute 1.00 L of the pH 4.00 solution to get a solution of pH 6.00? To what volume would you dilute 1.00 L of the pH 4.00 solution to get a solution of pH 8.00?

**22.** A solution of perchloric acid, $HClO_3$, a strong acid, has a pH of 1.28. How many moles of NaOH would be required to react completely with the $HClO_3$ in 1.00 L of the solution? What mass of NaOH is required?

**23.** A solution of the weak base $NH_3$ has a pH of 11.90. How many moles of HCl would have to be added to 1.00 L of the ammonia to react with all of the $OH^-$ ions present at pH 11.90?

**24.** The pH of a citric acid solution is 3.15. What are the $[H_3O^+]$ and $[OH^-]$ in this solution?

Skills Worksheet

# Problem Solving

## Titrations

Chemists have many methods for determining the quantity of a substance present in a solution or other mixture. One common method is titration, in which a solution of known concentration reacts with a sample containing the substance of unknown quantity. There are two main requirements for making titration possible. Both substances must react quickly and completely with each other, and there must be a way of knowing when the substances have reacted in precise stoichiometric quantities.

The most common titrations are acid-base titrations. These reactions are easily monitored by keeping track of pH changes with a pH meter or by choosing an indicator that changes color when the acid and base have reacted in stoichiometric quantities. This point is referred to as the *equivalence point*. Look at the following equation for the neutralization of KOH with HCl.

$$KOH(aq) + HCl(aq) \rightarrow KCl(aq) + H_2O(l)$$

Suppose you have a solution that contains 1.000 mol of KOH. All of the KOH will have reacted when 1.000 mol of HCl has been added. This is the equivalence point of this reaction.

Titration calculations rely on the relationship between volume, concentration, and amount.

*volume of solution* $\times$ *molarity of solution = amount of solute in moles*

If a titration were carried out between KOH and HCl, according the reaction above, the amount in moles of KOH and HCl would be equal at the equivalence point. The following relationship applies to this system:

*molarity*$_{KOH}$ $\times$ *volume*$_{KOH}$ = *amount of KOH in moles*

*amount of KOH in moles = amount of HCl in moles*

*amount of HCl in moles = molarity*$_{HCl}$ $\times$ *volume*$_{HCl}$

Therefore:

*molarity*$_{KOH}$ $\times$ *volume*$_{KOH}$ = *molarity*$_{HCl}$ $\times$ *volume*$_{HCl}$

The following plan for solving titration problems may be applied to any acid-base titration, regardless of whether the equivalence point occurs at equivalent volumes.

## General Plan for Solving Titration Problems

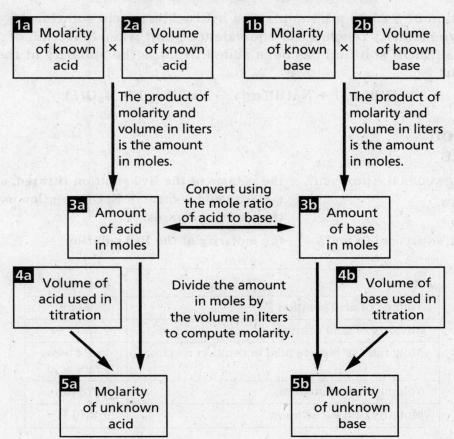

| 1a Molarity of known acid | × | 2a Volume of known acid | | 1b Molarity of known base | × | 2b Volume of known base |

The product of molarity and volume in liters is the amount in moles.

The product of molarity and volume in liters is the amount in moles.

| 3a Amount of acid in moles | Convert using the mole ratio of acid to base. | 3b Amount of base in moles |

| 4a Volume of acid used in titration | Divide the amount in moles by the volume in liters to compute molarity. | 4b Volume of base used in titration |

| 5a Molarity of unknown acid | | 5b Molarity of unknown base |

**▌Problem Solving** *continued*

## Sample Problem 1

A titration of a 25.00 mL sample of a hydrochloric acid solution of unknown molarity reaches the equivalence point when 38.28 mL of 0.4370 M NaOH solution has been added. What is the molarity of the HCl solution?

$$HCl(aq) + NaOH(aq) \rightarrow NaCl(aq) + H_2O(l)$$

## Solution

### ANALYZE

What is given in the problem?    the volume of the HCl solution titrated, and the molarity and volume of NaOH solution used in the titration figures.

What are you asked to find?    the molarity of the HCl solution

| Items | Data |
|---|---|
| Volume of acid solution | 25.00 mL |
| Molarity of acid solution | ? M |
| Mole ratio of base to acid in titration reaction | 1 mol base: 1 mol acid |
| Volume of base solution | 38.28 mL |
| Molarity of base solution | 0.4370 M |

### PLAN

What steps are needed to calculate the molarity of the HCl solution?

Use the volume and molarity of the NaOH to calculate the number of moles of NaOH that reacted. Use the mole ratio between base and acid to determine the moles of HCl that reacted. Use the volume of the acid to calculate molarity.

**▌Problem Solving** *continued*

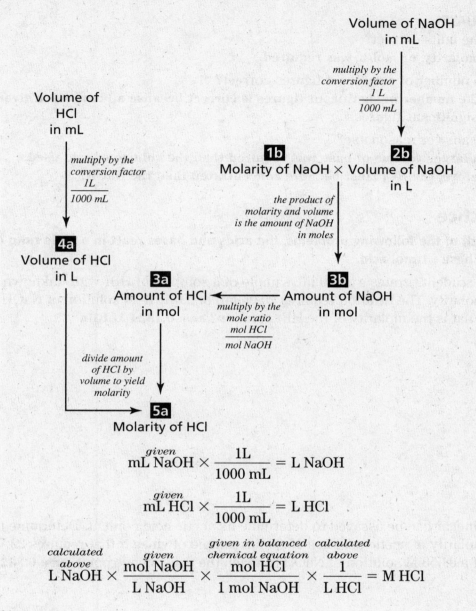

Volume of NaOH
in mL

*multiply by the
conversion factor*
$\dfrac{1\ L}{1000\ mL}$

Volume of
HCl
in mL

*multiply by the
conversion factor*
$\dfrac{1L}{1000\ mL}$

**1b**
Molarity of NaOH × Volume of NaOH
in L

**2b**

*the product of
molarity and volume
is the amount of NaOH
in moles*

**4a**
Volume of HCl
in L

**3a**
Amount of HCl
in mol

*multiply by the
mole ratio*
$\dfrac{mol\ HCl}{mol\ NaOH}$

**3b**
Amount of NaOH
in mol

*divide amount
of HCl by
volume to yield
molarity*

**5a**
Molarity of HCl

$$\text{mL NaOH} \overset{given}{\times} \frac{1L}{1000\ mL} = \text{L NaOH}$$

$$\text{mL HCl} \overset{given}{\times} \frac{1L}{1000\ mL} = \text{L HCl}$$

$$\underset{above}{\overset{calculated}{\text{L NaOH}}} \times \underset{\text{L NaOH}}{\overset{given}{\frac{\text{mol NaOH}}{}}} \times \underset{\text{1 mol NaOH}}{\overset{given\ in\ balanced\ chemical\ equation}{\frac{\text{mol HCl}}{}}} \times \underset{\text{L HCl}}{\overset{calculated\ above}{\frac{1}{}}} = \text{M HCl}$$

**COMPUTE**

$$38.28\ \text{mL NaOH} \times \frac{1L}{1000\ mL} = 0.03828\ \text{L NaOH}$$

$$25.00\ \text{mL HCl} \times \frac{1L}{1000\ mL} = 0.02500\ \text{L HCl}$$

$$0.03828\ \text{L NaOH} \times \frac{0.4370\ \text{mol NaOH}}{\text{L NaOH}} \times \frac{1\ \text{mol HCl}}{1\ \text{mol NaOH}}$$

$$\times \frac{1}{0.02500\ \text{L HCl}} = 0.6691\ \text{M HCl}$$

## EVALUATE

Are the units correct?

**Yes; molarity, or mol/L, was required.**

Is the number of significant figures correct?

**Yes; the number of significant figures is correct because all data were given to four significant figures.**

Is the answer reasonable?

**Yes; a larger volume of base was required than the volume of acid used. Therefore, the HCl must be more concentrated than the NaOH.**

# Practice

**In each of the following problems, the acids and bases react in a mole ratio of 1mol base : 1 mol acid.**

1. A student titrates a 20.00 mL sample of a solution of HBr with unknown molarity. The titration requires 20.05 mL of a 0.1819 M solution of NaOH. What is the molarity of the HBr solution? **ans: 0.1824 M HBr**

2. Vinegar can be assayed to determine its acetic acid content. Determine the molarity of acetic acid in a 15.00 mL sample of vinegar that requires 22.70 mL of a 0.550 M solution of NaOH to reach the equivalence point. **ans: 0.832 M**

# Sample Problem 2

A 50.00 mL sample of a sodium hydroxide solution is titrated with a
1.605 M solution of sulfuric acid. The titration requires 24.09 mL of the
acid solution to reach the equivalence point. What is the molarity of the
base solution?

$$H_2SO_4(aq) + 2NaOH(aq) \rightarrow Na_2SO_4(aq) + 2H_2O(l)$$

# Solution

## ANALYZE

What is given in the problem?    the balanced chemical equation for the acid-
base reaction, the volume of the base solution,
and the molarity and volume of the acid used in
the titration

What are you asked to find?    the molarity of the sodium hydroxide solution

| Items | Data |
|---|---|
| Volume of acid solution | 24.09 mL |
| Molarity of acid solution | 1.605 M |
| Mole ratio of base to acid in titration reaction | 2 mol base: 1 mol acid |
| Volume of base solution | 50.00 mL |
| Molarity of base solution | ? M |

## PLAN

What steps are needed to calculate the molarity of the NaOH solution?
Use the volume and molarity of the acid to calculate the number of moles of acid
that reacted. Use the mole ratio between base and acid to determine the moles
of base that reacted. Use the volume of the base to calculate molarity.

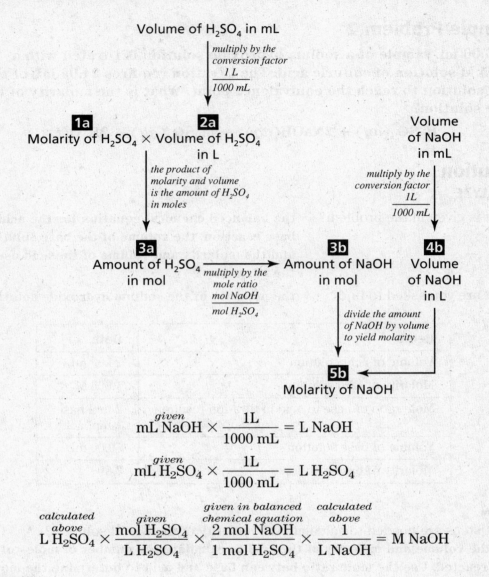

Volume of $H_2SO_4$ in mL

*multiply by the conversion factor*

$$\frac{1\ L}{1000\ mL}$$

**1a**
Molarity of $H_2SO_4$ × **2a** Volume of $H_2SO_4$ in L

*the product of molarity and volume is the amount of $H_2SO_4$ in moles*

Volume of NaOH in mL

*multiply by the conversion factor*

$$\frac{1L}{1000\ mL}$$

**3a**
Amount of $H_2SO_4$ in mol

*multiply by the mole ratio*

$$\frac{mol\ NaOH}{mol\ H_2SO_4}$$

**3b** Amount of NaOH in mol **4b** Volume of NaOH in L

*divide the amount of NaOH by volume to yield molarity*

**5b**
Molarity of NaOH

$$\overset{given}{mL\ NaOH} \times \frac{1L}{1000\ mL} = L\ NaOH$$

$$\overset{given}{mL\ H_2SO_4} \times \frac{1L}{1000\ mL} = L\ H_2SO_4$$

$$\overset{\substack{calculated \\ above}}{L\ H_2SO_4} \times \overset{given}{\frac{mol\ H_2SO_4}{L\ H_2SO_4}} \times \overset{\substack{given\ in\ balanced \\ chemical\ equation}}{\frac{2\ mol\ NaOH}{1\ mol\ H_2SO_4}} \times \overset{\substack{calculated \\ above}}{\frac{1}{L\ NaOH}} = M\ NaOH$$

**COMPUTE**

$$50.00\ mL\ NaOH \times \frac{1L}{1000\ mL} = 0.05000\ L\ NaOH$$

$$24.09\ mL\ H_2SO_4 \times \frac{1L}{1000\ mL} = 0.02409\ L\ H_2SO_4$$

$$0.02409\ L\ H_2SO_4 \times \frac{1.605\ mol\ H_2SO_4}{L\ H_2SO_4} \times \frac{2\ mol\ NaOH}{1\ mol\ H_2SO_4}$$

$$\times \frac{1}{0.05000\ L\ NaOH} = 1.547\ M\ NaOH$$

**▌Problem Solving** *continued*

**EVALUATE**

Are the units correct?

**Yes; molarity, or mol/L, was required.**

Is the number of significant figures correct?

**Yes; the number of significant figures is correct because all data were given to four significant figures.**

Is the answer reasonable?

**Yes; the volume of acid required was approximately half the volume of base used. Because of the 1:2 mole ratio, the acid must be about the same as the concentration of the base, which agrees with the result obtained.**

# Practice

**1.** A 20.00 mL sample of a solution of $Sr(OH)_2$ is titrated to the equivalence point with 43.03 mL of 0.1159 M HCl. What is the molarity of the $Sr(OH)_2$ solution? **ans: 0.1247 M $Sr(OH)_2$**

**2.** A 35.00 mL sample of ammonia solution is titrated to the equivalence point with 54.95 mL of a 0.400 M sulfuric acid solution. What is the molarity of the ammonia solution? **ans: 1.26 M $NH_3$**

# Sample Problem 3

A supply of NaOH is known to contain the contaminants NaCl and MgCl$_2$. A 4.955 g sample of this material is dissolved and diluted to 500.00 mL with water. A 20.00 mL sample of this solution is titrated with 22.26 mL of a 0.1989 M solution of HCl. What percentage of the original sample is NaOH? Assume that none of the contaminants react with HCl.

## Solution

### ANALYZE

What is given in the problem?    the mass of the original solute sample, the volume of the solution of the sample, the volume of the sample taken for titration, the molarity of the acid solution, and the volume of the acid solution used in the titration

What are you asked to find?    the percentage by mass of NaOH in the original sample

| Items | Data |
|---|---|
| Volume of acid solution | 22.26 mL |
| Molarity of acid solution | 0.1989 M |
| Mole ratio of base to acid in titration reaction | ? |
| Volume of base solution titrated | 20.00 mL |
| Moles of base in solution titrated | ? mol NaOH |
| Volume of original sample solution | 500.00 mL |
| Moles of base in original sample | ? mol NaOH |
| Mass of original sample | 4.955 g impure NaOH |
| Mass of base in original sample | ? g NaOH |
| Percentage of NaOH in original sample | ?% NaOH |

### PLAN

What steps are needed to calculate the concentration of NaOH in the sample?
Determine the balanced chemical equation for the titration reaction. Use the volume and molarity of the HCl to calculate the moles of HCl that reacted. Use the mole ratio between base and acid to determine the amount of NaOH that reacted. Divide by the volume titrated to obtain the concentration of NaOH.

## Problem Solving continued

What steps are needed to calculate the percentage of NaOH in the sample?
**Convert the concentration of NaOH to the amount of NaOH in the original
sample by multiplying the concentration by the total volume. Convert amount of
NaOH to mass NaOH by using the molar mass of NaOH. Use the mass of NaOH
and the mass of the sample to calculate the percentage of NaOH.**

You must first determine the equation for titration reaction.

$$HCl(aq) + NaOH(aq) \rightarrow NaCl(aq) + H_2O(l)$$

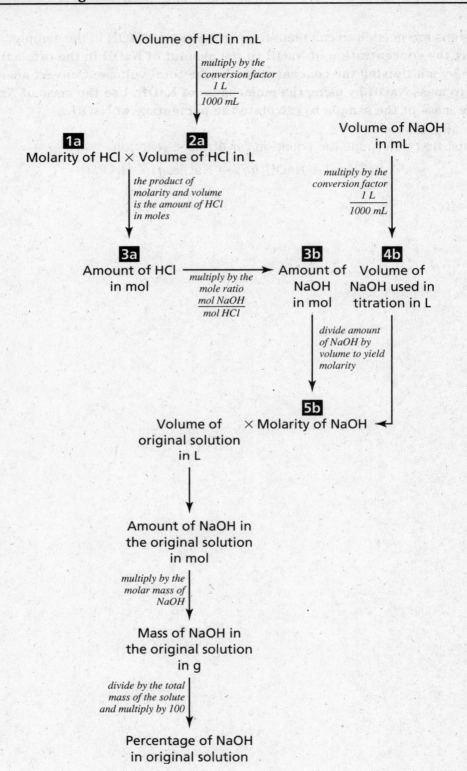

Volume of HCl in mL

*multiply by the conversion factor*
$$\frac{1\ L}{1000\ mL}$$

**1a**
Molarity of HCl ×

**2a**
Volume of HCl in L

*the product of molarity and volume is the amount of HCl in moles*

Volume of NaOH in mL

*multiply by the conversion factor*
$$\frac{1\ L}{1000\ mL}$$

**3a**
Amount of HCl in mol

*multiply by the mole ratio*
$$\frac{mol\ NaOH}{mol\ HCl}$$

**3b**
Amount of NaOH in mol

**4b**
Volume of NaOH used in titration in L

*divide amount of NaOH by volume to yield molarity*

**5b**
× Molarity of NaOH

Volume of original solution in L

Amount of NaOH in the original solution in mol

*multiply by the molar mass of NaOH*

Mass of NaOH in the original solution in g

*divide by the total mass of the solute and multiply by 100*

Percentage of NaOH in original solution

**| Problem Solving** continued

$$\text{mL NaOH}_{titrated}^{given} \times \frac{1L}{1000 \text{ mL}} = \text{L NaOH}_{titrated}$$

$$\text{mL HCl}^{given} \times \frac{1L}{1000 \text{ mL}} = \text{L HCl}$$

$$\text{L HCl}^{\substack{calculated \\ above}} \times \frac{\text{mol HCl}^{given}}{\text{L HCl}} \times \frac{1 \text{ mol NaOH}^{\substack{given \ in \ balanced \\ chemical \ equation}}}{1 \text{ mol HCl}} \times \frac{1}{\text{L NaOH}_{titrated}} = \text{M NaOH}$$

$$\text{mL NaOH}_{original}^{given} \times \frac{1L}{1000 \text{ mL}} = \text{L NaOH}_{original}$$

$$\frac{\text{mol NaOH}^{\substack{calculated \\ above}}}{\text{L NaOH}} \times \text{L NaOH}_{original}^{\substack{calculated \\ above}} \times \frac{40.00 \text{ g NaOH}^{\substack{molar \ mass \\ of \ NaOH}}}{\text{mol NaOH}} = \text{g NaOH}_{original}$$

$$\frac{\text{g NaOH}_{original}^{\substack{calculated \\ above}}}{\text{g solute}_{given}} \times 100 = \text{percentage of NaOH in solute}$$

## COMPUTE

$$20.00 \text{ mL NaOH}_{titrated} \times \frac{1L}{1000 \text{ mL}} = 0.02000 \text{ L NaOH}_{titrated}$$

$$22.26 \text{ mL HCl} \times \frac{1 \text{ L}}{1000 \text{ mL}} = 0.02226 \text{ L HCl}$$

$$0.02226 \text{ L HCl} \times \frac{0.1989 \text{ mol HCl}}{\text{L HCl}} \times \frac{1 \text{ mol NaOH}}{1 \text{ mol HCl}}$$

$$\times \frac{1}{0.02000 \text{ L NaOH}} = 0.2214 \text{ M NaOH}$$

$$500.00 \text{ mL NaOH}_{original} \times \frac{1 \text{ L}}{1000 \text{ mL}} = 0.500 \ 00 \text{ L NaOH}_{original}$$

$$\frac{0.2214 \text{ mol NaOH}}{\text{L NaOH}} \times 0.500 \ 00 \text{ L NaOH} \times \frac{40.00 \text{ g NaOH}}{\text{mol NaOH}} = 4.428 \text{ g NaOH}$$

$$\frac{4.428 \text{ g NaOH}}{4.955 \text{ g solute}} \times 100 = 89.35\% \text{ NaOH}$$

## EVALUATE
Are the units correct?
**Yes; units canceled to give percentage of NaOH in sample.**

Is the number of significant figures correct?

**Yes; the number of significant figures is correct because all data were given to four significant figures.**

Is the answer reasonable?

**Yes; the calculation can be approximated as $(0.02 \times 0.2 \times 25 \times 40 \times 100)/5 = 400/5 = 80$, which is close to the calculated result.**

## Practice

**In the problems below, assume that impurities are not acidic or basic and that they do not react in an acid-base titration.**

1. A supply of glacial acetic acid has absorbed water from the air. It must be assayed to determine the actual percentage of acetic acid. 2.000 g of the acid is diluted to 100.00 mL, and 20.00 mL is titrated with a solution of sodium hydroxide. The base solution has a concentration of 0.218 M, and 28.25 mL is used in the titration. Calculate the percentage of acetic acid in the original sample. Write the titration equation to get the mole ratio.
   **ans: 92.5% acetic acid**

2. A shipment of crude sodium carbonate must be assayed for its $Na_2CO_3$ content. You receive a small jar containing a sample from the shipment and weigh out 9.709 g into a flask, where it is dissolved in water and diluted to 1.0000 L with distilled water. A 10.00 mL sample is taken from the flask and titrated to the equivalence point with 16.90 mL of a 0.1022 M HCl solution. Determine the percentage of $Na_2CO_3$ in the sample. Write the titration equation to get the mole ratio. **ans: 94.28% $Na_2CO_3$**

| Problem Solving *continued*

# Additional Problems—Titrations

1. A 50.00 mL sample of a potassium hydroxide is titrated with a 0.8186 M HCl solution. The titration requires 27.87 mL of the HCl solution to reach the equivalence point. What is the molarity of the KOH solution?

2. A 15.00 mL sample of acetic acid is titrated with 34.13 mL of 0.9940 M NaOH. Determine the molarity of the acetic acid.

3. A 12.00 mL sample of an ammonia solution is titrated with 1.499 M $HNO_3$ solution. A total of 19.48 mL of acid is required to reach the equivalence point. What is the molarity of the ammonia solution?

4. A certain acid and base react in a 1:1 ratio.

   **a.** If the acid and base solutions are of equal concentration, what volume of acid will titrate a 20.00 mL sample of the base?

   **b.** If the acid is twice as concentrated as the base, what volume of acid will be required to titrate 20.00 mL of the base?

   **c.** How much acid will be required if the base is four times as concentrated as the acid, and 20.00 mL of base is used?

5. A 10.00 mL sample of a solution of hydrofluoric acid, HF, is diluted to 500.00 mL. A 20.00 mL sample of the diluted solution requires 13.51 mL of a 0.1500 M NaOH solution to be titrated to the equivalence point. What is the molarity of the original HF solution?

6. A solution of oxalic acid, a diprotic acid, is used to titrate a 16.22 mL sample of a 0.5030 M KOH solution. If the titration requires 18.41 mL of the oxalic acid solution, what is its molarity?

7. A $H_2SO_4$ solution of unknown molarity is titrated with a 1.209 M NaOH solution. The titration requires 42.27 mL of the NaOH solution to reach the equivalent point with 25.00 mL of the $H_2SO_4$ solution. What is the molarity of the acid solution?

8. Potassium hydrogen phthalate, $KHC_8H_4O_4$, is a solid acidic substance that reacts in a 1:1 mole ratio with bases that have one hydroxide ion. Suppose that 0.7025 g of $KHC_8H_4O_4$ is titrated to the equivalence point by 20.18 mL of a KOH solution. What is the molarity of the KOH solution?

9. A solution of citric acid, a triprotic acid, is titrated with a sodium hydroxide solution. A 20.00 mL sample of the citric acid solution requires 17.03 mL of a 2.025 M solution of NaOH to reach the equivalence point. What is the molarity of the acid solution?

10. A flask contains 41.04 mL of a solution of potassium hydroxide. The solution is titrated and reaches an equivalence point when 21.65 mL of a 0.6515 M solution of $HNO_3$ is added. Calculate the molarity of the base solution.

11. A bottle is labeled 2.00 M $H_2SO_4$. You decide to titrate a 20.00 mL sample with a 1.85 M NaOH solution. What volume of NaOH solution would you expect to use if the label is correct?

**| Problem Solving** *continued*

---

**12.** What volume of a 0.5200 M solution of $H_2SO_4$ would be needed to titrate 100.00 mL of a 0.1225 M solution of $Sr(OH)_2$?

**13.** A sample of a crude grade of KOH is sent to the lab to be tested for KOH content. A 4.005 g sample is dissolved and diluted to 200.00 mL with water. A 25.00 mL sample of the solution is titrated with a 0.4388 M HCl solution and requires 19.93 mL to reach the equivalence point. How many moles of KOH were in the 4.005 g sample? What mass of KOH is this? What is the percent KOH in the crude material?

**14.** What mass of magnesium hydroxide would be required for the magnesium hydroxide to react to the equivalence point with 558 mL of 3.18 M hydrochloric acid?

**15.** An ammonia solution of unknown concentration is titrated with a solution of hydrochloric acid. The HCl solution is 1.25 M, and 5.19 mL are required to titrate 12.61 mL of the ammonia solution. What is the molarity of the ammonia solution?

**16.** What volume of 2.811 M oxalic acid solution is needed to react to the equivalence point with a 5.090 g sample of material that is 92.10% NaOH? Oxalic acid is a diprotic acid.

**17.** Standard solutions of accurately known concentration are available in most laboratories. These solutions are used to titrate other solutions to determine their concentrations. Once the concentration of the other solutions are accurately known, they may be used to titrate solutions of unknowns.

The molarity of a solution of HCl is determined by titrating the solution with an accurately known solution of $Ba(OH)_2$, which has a molar concentration of 0.1529 M. A volume of 43.09 mL of the $Ba(OH)_2$ solution titrates 26.06 mL of the acid solution. The acid solution is in turn used to titrate 15.00 mL of a solution of rubidium hydroxide. The titration requires 27.05 mL of the acid.

**a.** What is the molarity of the HCl solution?

**b.** What is the molarity of the RbOH solution?

**18.** A truck containing 2800 kg of a 6.0 M hydrochloric acid has been in an accident and is in danger of spilling its load. What mass of $Ca(OH)_2$ should be sent to the scene in order to neutralize all of the acid in case the tank bursts? The density of the 6.0 M HCl solution is 1.10 g/mL.

**19.** A 1.00 mL sample of a fairly concentrated nitric acid solution is diluted to 200.00 mL. A 10.00 mL sample of the diluted solution requires 23.94 mL of a 0.0177 M solution of $Ba(OH)_2$ to be titrated to the equivalence point. Determine the molarity of the original nitric acid solution.

**20.** What volume of 4.494 M $H_2SO_4$ solution would be required to react to the equivalence point with 7.2280 g of $LiOH(s)$?

Skills Worksheet

# Problem Solving

## Equilibrium of Acids and Bases, $K_a$ and $K_b$

The ionization of a strong acid, such as HCl, will proceed to completion in reasonably dilute solutions. This process is written as follows.

$$HCl(g) + H_2O(l) \rightarrow H_3O^+(aq) + Cl^-(aq)$$
*goes 100% to the right; NOT an equilibrium process.*

However, when weak acids and weak bases dissolve, they only partially ionize, resulting in an equilibrium between ionic and molecular forms. The following equation shows the equilibrium process that occurs when hydrogen fluoride, a weak acid, dissolves in water.

$$HF(g) + H_2O(l) \rightleftharpoons H_3O^+(aq) + F^-(aq)$$
*does not go 100% to the right; IS an equilibrium process*

Most weak acids react in this way, that is, by donating a proton to a water molecule to form a $H_3O^+$ ion.

The weak base ammonia establishes the following equilibrium in water.

$$NH_3(g) + H_2O(l) \rightleftharpoons NH_4^+(aq) + OH^-(aq)$$

Most weak bases react in this way, that is, by accepting a proton from a water molecule to leave an $OH^-$ ion.

You learned to write an equilibrium expression to solve for $K$, the equilibrium constant of a reaction. As with equilibrium reactions, equilibrium constants can be calculated for the ionization and dissociation processes shown above. The equilibrium constants indicate how far the equilibrium goes toward the ionic "products." Recall that the concentration of each reaction component is raised to the power of its coefficient in the balanced equation. These concentration terms are arranged in a fraction with products in the numerator and reactants in the denominator. The equilibrium constants calculated for acid ionization reactions are called *acid ionization constants* and have the symbol $K_a$. The equilibrium constants calculated for base dissociation reactions are called *base dissociation constants* and have the symbol $K_b$. From the diagram on the next page and the problems in this chapter, you will see how these constants are calculated and how they relate to concentrations of the reactants and products and to pH.

**Problem Solving** *continued*

## General Plan for Solving $K_a$ and $K_b$ Problems

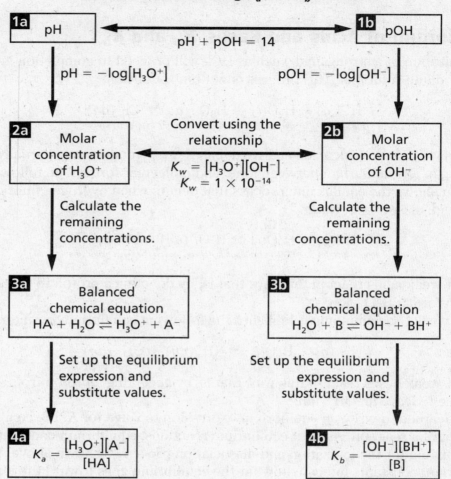

# Sample Problem 1

The hydronium ion concentration of a 0.500 M solution of HF at 25°C is found to be 0.0185 M. Calculate the ionization constant for HF at this temperature.

# Solution

## ANALYZE

What is given in the problem?    **the molarity of the acid solution, the equilibrium concentration of hydronium ions, and the temperature**

What are you asked to find?    **the acid ionization constant, $K_a$**

| Items | Data |
|---|---|
| Molar concentration of $H_3O^+$ at equilibrium | 0.0185 M |
| Molar concentration of $F^-$ at equilibrium | ? M |
| Initial molar concentration of HF | 0.500 M |
| Molar concentration of HF at equilibrium | ? M |
| Acid ionization constant, $K_a$, of HF | ? |

## PLAN

What steps are needed to calculate the acid ionization constant for HF?
Write the equation for the ionization reaction. Set up the equilibrium expression for the ionization of HF in water. Determine the concentrations of all components at equilbrium, and calculate $K_a$.

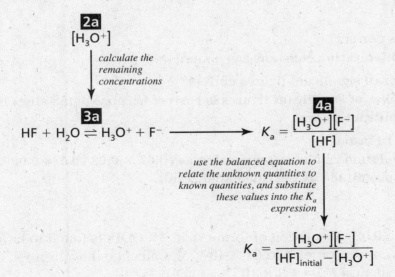

$$K_a = \frac{[H_3O^+][F^-]}{[HF]_{initial} - [H_3O^+]}$$

| **Problem Solving** *continued* |

Write the balanced chemical equation for the ionization of HF in aqueous solution.

$$HF(aq) + H_2O(l) \rightarrow H_3O^+(aq) + F^-(aq)$$

Write the acid ionization expression for $K_a$. Remember that pure substances are not included in equilibrium expressions. For this reason, $[H_2O]$ does not appear in the expression for $K_a$.

$$K_a = \frac{[H_3O^+][F^-]}{[HF]}$$

From the balanced chemical equation, you can see that one HF molecule ionizes to give one fluoride ion and one hydronium ion. Therefore, at equilibrium, the concentration of $F^-$ must equal the concentration of $H_3O^+$.

$$[F^-] = [H_3O^+] = 0.0185 \text{ M}$$

When the HF ionizes in solution, the HF concentration decreases from its initial value. The amount by which it decreases is equal to the concentration of either the fluoride ion or the hydronium ion.

$$[HF]_{equilibrium} = [HF]_{initial} - [H_3O^+]_{equilibrium}$$

Set up the equilibrium expression.

$$K_a = \frac{\overset{given}{[H_3O^+]}\,\overset{\overset{calculated}{above}}{[F^-]}}{\underset{given}{[HF]_{initial}} - \underset{given}{[H_3O^+]}}$$

## COMPUTE

$$K_a = \frac{[0.0185][0.0185]}{[0.500] - [0.0185]} = 7.11 \times 10^{-4}$$

## EVALUATE

Are the units correct?

**Yes; the acid ionization constant has no units.**

Is the number of significant figures correct?

**Yes; the number of significant figures is correct because data values have as few as three significant figures.**

Is the answer reasonable?

**Yes; the calculation can be approximated as $(0.02 \times 0.02)/0.5 = 0.0008$, which is of the same magnitude as the calculated result.**

## PRACTICE

**1.** At 25°C, a 0.025 M solution of formic acid, HCOOH, is found to have a hydronium ion concentration of $2.03 \times 10^{-3}$ M. Calculate the ionization constant of formic acid. **ans: $K_a = 1.8 \times 10^{-4}$**

**| Problem Solving** *continued*

# Sample Problem 2

At 25°C, the pH of a 0.315 M solution of nitrous acid, $HNO_2$, is 1.93.
Calculate the $K_a$ of nitrous acid at this temperature.

# Solution

## ANALYZE

What is given in the problem?     **the pH of the acid solution and the original**
                                  **concentration of $HNO_2$**

What are you asked to find?       **the acid ionization constant, $K_a$**

| Items | Data |
|---|---|
| pH of solution | 1.93 |
| Molar concentration of $H_3O^+$ at equilibrium | ? M |
| Molar concentration of $NO_2^-$ at equilibrium | ? M |
| Initial molar concentration of $HNO_2$ | 0.315 M |
| Molar concentration of $HNO_2$ at equilibrium | ? M |
| Acid ionization constant, $K_a$, of $HNO_2$ | ? |

## PLAN

What steps are needed to calculate the acid ionization constant for $HNO_2$?
**Determine the $H_3O^+$ concentration from the pH. Write the equation for the
ionization reaction. Set up the equilibrium expression. Determine all equilibrium concentrations and calculate $K_a$.**

**1a**
$$pH = -\log[H_3O^+]$$

*rearrange to solve
for $[H_3O^+]$*

**2a**
$$[H_3O^+] = 10^{-pH}$$

**3a**
$$HNO_2 + H_2O \rightleftharpoons H_3O^+ + NO_2^-$$

**4a**
$$K_a = \frac{[H_3O^+][NO_2^-]}{[HNO_2]}$$

*use the balanced equation to
relate the unknown quantities to
known quantities, and substitute
these values into the $K_a$
expression*

$$K_a = \frac{[H_3O^+][NO_2^-]}{[HNO_2]_{initial} - [H_3O^+]}$$

**| Problem Solving** *continued*

Calculate the $H_3O^+$ concentration from the pH.

$$pH = -\log [H_3O^+]$$

$$[H_3O^+] = 10^{\overset{given}{-pH}}$$

Write the balanced chemical equation for the ionization of $HNO_2$ in aqueous solution.

$$HNO_2(aq) + H_2O(l) \rightleftarrows H_3O^+(aq) + NO_2^-(aq)$$

Write the mathematical equation to compute $K_a$.

$$K_a = \frac{[NO_2^-][H_3O^+]}{[HNO_2]}$$

Because each $HNO_2$ molecule dissociates into one hydronium ion and one nitrite ion,

$$[NO_2^-] = [H_3O^+]$$

the $HNO_2$ concentration at equilibrium will be its initial concentration minus any $HNO_2$ that has ionized. The amount ionized will equal the concentration of $H_3O^+$ or $NO_2^-$.

$$[HNO_2]_{equilibrium} = [HNO_2]_{initial} - [H_3O^+]$$

Substitute known values into the $K_a$ expression.

$$K_a = \frac{\overset{\substack{calculated \\ above}}{[H_3O^+]} \, \overset{\substack{calculated \\ above}}{[NO_2^-]}}{\underset{given}{[HNO_2]_{initial}} - \underset{\substack{calculated \\ above}}{[H_3O^+]}}$$

## COMPUTE

$$[H_3O^+] = 10^{-1.93} = 1.2 \times 10^{-2}$$

$$K_a = \frac{[1.2 \times 10^{-2}][1.2 \times 10^{-2}]}{[0.315] - [1.2 \times 10^{-2}]} = 4.4 \times 10^{-4}$$

## EVALUATE

Are the units correct?

**Yes; the acid ionization constant has no units.**

Is the number of significant figures correct?

**Yes; the number of significant figures is correct because the pH was given to two decimal places.**

Is the answer reasonable?

**Yes; the calculation can be approximated as $(0.01 \times 0.01)/0.3 = 3 \times 10^{-4}$, which is of the same order of magnitude as the calculated result.**

**Problem Solving** *continued*

## Practice

**1.** The pH of a 0.400 M solution of iodic acid, $HIO_3$, is 0.726 at 25°C. What is the $K_a$ at this temperature? **ans: $K_a$ = 0.167**

**2.** The pH of a 0.150 M solution of hypochlorous acid, HClO, is found to be 4.55 at 25°C. Calculate the $K_a$ for HClO at this temperature. **ans: $K_a$ = 5.2 × $10^{-9}$**

## Sample Problem 3

A 0.450 M ammonia solution has a pH of 11.45 at 25°C. Calculate the $[H_3O^+]$ and $[OH^-]$ of the solution, and determine the base dissociation constant, $K_b$, of ammonia.

## Solution

### ANALYZE

What is given in the problem?     the pH of the base solution, the original concentration of $NH_3$, and the temperature

What are you asked to find?     $[H_3O^+]$, $[OH^-]$, and the base dissociation constant, $K_b$

| Items | Data |
|---|---|
| pH of solution | 11.45 |
| Molar concentration of $H_3O^+$ at equilibrium | ? M |
| Molar concentration of $OH^-$ at equilibrium | ? M |
| Molar concentration of $NH_4^+$ at equilibrium | ? M |
| Initial molar concentration of $NH_3$ | 0.450 M |
| Molar concentration of $NH_3$ at equilibrium | ? M |
| Base dissociation constant, $K_b$, of $NH_3$ | ? |

### PLAN

What steps are needed to calculate the base dissociation constant for $NH_3$? Determine the $H_3O^+$ concentration from the pH. Calculate the $OH^-$ concentration. Write the balanced chemical equation for the dissociation reaction. Write the mathematical equation for $K_b$, substitute values, and calculate.

## Problem Solving *continued*

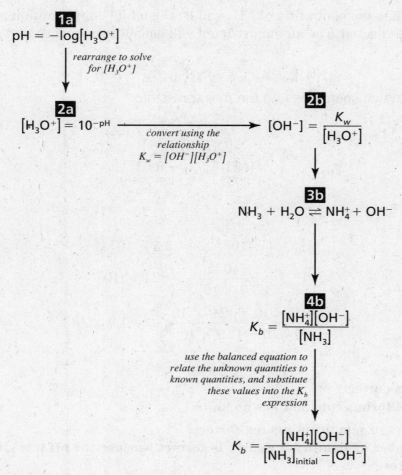

Determine $[H_3O^+]$.

$$[H_3O^+] = 10^{\overset{given}{-pH}}$$

Determine $[OH^-]$.

$$K_w = [H_3O^+][OH^-]$$

$$[OH^-] = \frac{K_w}{\underset{\underset{above}{calculated}}{[H_3O^+]}}$$

Write the equilibrium expression for the ionization reaction.

$$NH_3(aq) + H_2O(l) \rightleftarrows NH_4^+(aq) + OH^-(aq)$$

$$K_b = \frac{[OH^-][NH_4^+]}{[NH_3]}$$

Because 1 mol of $NH_3$ reacts with 1 mol of $H_2O$ to produce 1 mol of $NH_4^+$ and 1 mol of $OH^-$:

$$[OH^-] = [NH_4^+]$$

**▌Problem Solving** *continued*

The equilibrium concentration of $NH_3$ will be the initial concentration minus any $NH_3$ that has reacted. The amount reacted will equal the concentration of $NH_4^+$ or $OH^-$.

$$[NH_3]_{equilibrium} = [NH_3]_{initial} - [OH^-]$$

Substitute known quantities into the $K_b$ expression.

$$K_b = \frac{\overset{calculated}{\overset{above}{[OH^-][NH_4^+]}}}{\underset{given}{[NH_3]_{initial}} - \underset{\substack{calculated \\ above}}{[OH^-]}}$$

**COMPUTE**

$$[H_3O^+] = 10^{-11.45} = 3.5 \times 10^{-12}$$

$$[OH^-] = \frac{10^{-14}}{3.5 \times 10^{-12}} = 2.9 \times 10^{-3}$$

$$K_b = \frac{[2.9 \times 10^{-3}][2.9 \times 10^{-3}]}{[0.450] - [2.9 \times 10^{-3}]} = 1.9 \times 10^{-5}$$

**EVALUATE**

Are the units correct?

**Yes; the equilibrium constant has no units.**

Is the number of significant figures correct?

**Yes; the number of significant figures is correct because the pH was given to two decimal places.**

Is the answer reasonable?

**Yes. The calculation can be approximated as (0.003 × 0.003)/0.5 = 0.00009/5 = 1.8 × 10⁻⁵.**

# Practice

1. The compound propylamine, $CH_3CH_2CH_2NH_2$, is a weak base. At equilibrium, a 0.039 M solution of propylamine has an $OH^-$ concentration of $3.74 \times 10^{-3}$ M. Calculate the pH of this solution and $K_b$ for propylamine.
   **ans: pH = 11.573; $K_b$ = 4.0 × 10⁻⁴**

**Problem Solving** *continued*

# Sample Problem 4

A 1.00 M iodic acid, $HIO_3$, solution has an acid ionization constant of 0.169 at 25°C. Calculate the hydronium ion concentration of the solution at this temperature.

# Solution

## ANALYZE

What is given in the problem?    **the original concentration of $HIO_3$, the temperature, and $K_a$**

What are you asked to find?    **$[H_3O^+]$**

| Items | Data |
|---|---|
| Molar concentration of $H_3O^+$ at equilibrium | ? M |
| Initial molar concentration of $HIO_3$ | 1.00 M |
| Molar concentration of $HIO_3$ at equilibrium | ? M |
| Acid ionization constant, $K_a$, of $HIO_3$ | 0.169 |

## PLAN

What steps are needed to calculate the acid ionization constant for $HIO_3$?
**Write the balanced chemical equation for the ionization reaction. Write the equation for $K_a$. Substitute $x$ for the unknown values. Rearrange the $K_a$ expression so that a quadratic equation remains. Substitute known values, and solve for $x$.**

Write the equilibrium equation for the ionization reaction.

$$HIO_3(aq) + H_2O(l) \rightleftharpoons H_3O^+(aq) + IO_3^-(aq)$$

Use the chemical equation to write an expression for $Ka$.

$$K_a = \frac{[H_3O^+][IO_3^-]}{[HIO_3]}$$

Since 1 mol of $HIO_3$ reacts with 1 mol of $H_2O$ to produce 1 mol of $H_3O^+$ and 1 mol of $IO_3^-$:

$$[H_3O^+] = [IO_3^-]$$

**Problem Solving** *continued*

The equilibrium concentration of $HIO_3$ will be the initial concentration minus any $HIO_3$ that has reacted. The amount reacted will equal the concentration of $H_3O^+$ or $IO_3^-$.

$$[HIO_3]_{equilibrium} = [HIO_3]_{initial} - [H_3O^-]$$

Because $[H_3O^+]$ is an unknown quantity, substitute the variable $x$ for it to solve.

$$[HIO_3]_{equilibrium} = [HIO_3]_{initial} - x$$

$$\overset{given}{K_a} = \frac{x^2}{\underset{given}{[HIO_3]_{initial}} - x}$$

**COMPUTE**

$$0.169 = \frac{x^2}{1.00 - x}$$

$$0.169(1.00 - x) = x^2$$

$$0.169 - 0.169x = x^2$$

Rearrange the above equation.

$$x^2 + 0.169x - 0.169 = 0$$

Notice that this equation fits the form for a general quadratic equation.

$$ax^2 + bx + c = 0$$

where $a = 1$, $b = 0.169$, $c = -0.169$

Use the formula for solving quadratic equations.

$$x = \frac{-b + \sqrt{b^2 - 4ac}}{2a}$$

Substitute the values given above into the quadratic equation, and solve for $x$.

$$x = \frac{-0.169 + \sqrt{(0.169)^2 - 4(1)(-0.169)}}{2(1)}$$

$$x = 0.335 \text{ M}$$

**EVALUATE**

Are the units correct?
**Yes; the value has units of mol/L, or M.**

Is the number of significant figures correct?
**Yes; the number of significant figures is correct because the data values had three significant figures.**

**▌Problem Solving** *continued*

Is the answer reasonable?
**Yes; substituting the calculated [H₃O⁺] back into the equation for $K_a$ yields the given value.**

# Practice

**1.** The $K_a$ of nitrous acid is $4.6 \times 10^{-4}$ at 25°C. Calculate the $[H_3O^+]$ of a 0.0450 M nitrous acid solution. **ans: $4.4 \times 10^{-3}$ M**

**Problem Solving** *continued*

# Additional Problems—Equilibrium of Acids and Bases

1. Hydrazoic acid, $HN_3$, is a weak acid. The $[H_3O^+]$ of a 0.102 M solution of hydrazoic acid is $1.39 \times 10^{-3}$ M. Determine the pH of this solution, and calculate $K_a$ at 25°C for $HN_3$.

2. Bromoacetic acid, $BrCH_2COOH$, is a moderately weak acid. A 0.200 M solution of bromoacetic acid has a $H_3O^+$ concentration of 0.0192 M. Determine the pH of this solution and the $K_a$ of bromoacetic acid at 25°C.

3. A base, B, dissociates in water according to the following equation:

$$B + H_2O \rightleftarrows BH^+ + OH^-$$

Complete the following table for base solutions with the characteristics given.

| Initial [B] | [B] at equilibrium | [OH$^-$] | $K_b$ | [H$_3$O$^-$] | pH |
|---|---|---|---|---|---|
| **a.** 0.400 M | NA | $2.70 \times 10^{-4}$ M | ? | ? M | ? |
| **b.** 0.005 50 M | ? M | $8.45 \times 10^{-4}$ M | ? | NA | ? |
| **c.** 0.0350 M | ? M | ? M | ? | ? M | 11.29 |
| **d.** ? M | 0.006 28 M | 0.000 92 M | ? | NA | ? |

4. The solubility of benzoic acid, $C_6H_5COOH$, in water at 25°C is 2.9 g/L. The pH of this saturated solution is 2.92. Determine $K_a$ at 25°C for benzoic acid. (Hint: first calculate the initial concentration of benzoic acid.)

5. A 0.006 50 M solution of ethanolamine, $H_2NCH_2CH_2OH$, has a pH of 10.64 at 25°C. Calculate the $K_b$ of ethanolamine. What concentration of undissociated ethanolamine remains at equilibrium?

6. The weak acid hydrogen selenide, $H_2Se$, has two hydrogen atoms that can form hydronium ions. The second ionization is so small that the concentration of the resulting $H_3O^+$ is insignificant. If the $[H_3O^+]$ of a 0.060 M solution of $H_2Se$ is $2.72 \times 10^{-3}$ M at 25°C, what is the $K_a$ of the first ionization?

7. Pyridine, $C_5H_5N$, is a very weak base. Its $K_b$ at 25°C is $1.78 \times 10^{-9}$. Calculate the $[OH^-]$ and pH of a 0.140 M solution. Assume that the concentration of pyridine at equilibrium is equal to its initial concentration because so little pyridine is dissociated.

8. A solution of a monoprotic acid, HA, at equilibrium is found to have a 0.0208 M concentration of nonionized acid. The pH of the acid solution is 2.17. Calculate the initial acid concentration and $K_a$ for this acid.

9. Pyruvic acid, $CH_3COCOOH$, is an important intermediate in the metabolism of carbohydrates in the cells of the body. A solution made by dissolving 438 mg of pyruvic acid in 10.00 mL of water is found to have a pH of 1.34 at 25°C. Calculate $K_a$ for pyruvic acid.

**10.** The $[H_3O^+]$ of a solution of acetoacetic acid, $CH_3COCH_2COOH$, is $4.38 \times 10^{-3}$ M at 25°C. The concentration of nonionized acid is 0.0731 M at equilibrium. Calculate $K_a$ for acetoacetic acid at 25°C.

**11.** The $K_a$ of 2-chloropropanoic acid, $CH_3CHClCOOH$, is $1.48 \times 10^{-3}$. Calculate the $[H_3O^+]$ and the pH of a 0.116 M solution of 2-chloropropionic acid. Let $x = [H_3O^+]$. The degree of ionization of the acid is too large to ignore. If your set up is correct, you will have a quadratic equation to solve.

**12.** Sulfuric acid ionizes in two steps in water solution. For the first ionization shown in the following equation, the $K_a$ is so large that in moderately dilute solution the ionization can be considered 100%.

$$H_2SO_4 + H_2O \rightarrow H_3O^+ + HSO_4^-$$

The second ionization is fairly strong, and $K_a = 1.3 \times 10^{-2}$.

$$HSO_4^- + H_2O \rightleftarrows H_3O^+ + SO_4^{2-}$$

Calculate the total $[H_3O^+]$ and pH of a 0.0788 M $H_2SO_4$ solution. Hint: If the first ionization is 100%, what will $[HSO_4^-]$ and $[H_3O^+]$ be? Remember to account for the already existing concentration of $H_3O^+$ in the second ionization. Let $x = [SO_4^{2-}]$.

**13.** The hydronium ion concentration of a 0.100 M solution of cyanic acid, HOCN, is found to be $5.74 \times 10^{-3}$ M at 25°C. Calculate the ionization constant of cyanic acid. What is the pH of this solution?

**14.** A solution of hydrogen cyanide, HCN, has a 0.025 M concentration. The cyanide ion concentration is found to be $3.16 \times 10^{-6}$ M.

**a.** What is the hydronium ion concentration of this solution?

**b.** What is the pH of this solution?

**c.** What is the concentration of nonionized HCN in the solution? Be sure to use the correct number of significant figures.

**d.** Calculate the ionization constant of HCN.

**e.** How would you characterize the strength of HCN as an acid?

**f.** Determine the $[H_3O^+]$ for a 0.085 M solution of HCN.

**15.** A 1.20 M solution of dichloroacetic acid, $CCl_2HCOOH$, at 25°C has a hydronium ion concentration of 0.182 M.

**a.** What is the pH of this solution?

**b.** What is the $K_a$ of dichloroacetic acid at 25°C?

**c.** What is the concentration of nonionized dichloroacetic acid in this solution?

**d.** What can you say about the strength of dichloroacetic acid?

**| Problem Solving** *continued*

**16.** Phenol, $C_6H_5OH$, is a very weak acid. The pH of a 0.215 M solution of phenol at 25°C is found to be 5.61. Calculate the $K_a$ for phenol.

**17.** A solution of the simplest amino acid, glycine ($NH_2CH_2COOH$), is prepared by dissolving 3.75 g in 250.0 mL of water at 25°C. The pH of this solution is found to be 0.890.

**a.** Calculate the molarity of the glycine solution.

**b.** Calculate the $K_a$ for glycine.

**18.** Trimethylamine, $(CH_3)_3N$, dissociates in water the same way that $NH_3$ does—by accepting a proton from a water molecule. The $[OH^-]$ of a 0.0750 M solution of trimethylamine at 25°C is $2.32 \times 10^{-3}$ M. Calculate the pH of this solution and the $K_b$ of trimethylamine.

**19.** Dimethylamine, $(CH_3)_2NH$, is a weak base similar to the trimethylamine in item 18. A $5.00 \times 10^{-3}$ M solution of dimethylamine has a pH of 11.20 at 25°C. Calculate the $K_b$ of dimethylamine. Compare this $K_b$ with the $K_b$ for trimethylamine that you calculated in item 18. Which substance is the stronger base?

**20.** Hydrazine dissociates in water solution according to the following equations:

$$H_2NNH_2 + H_2O(l) \rightleftharpoons NH_2NNH_3^+(aq) + OH^-(aq)$$

$$H_2NNH_3^+(aq) + H_2O(l) \rightleftharpoons NH_3NNH_3^{2+}(aq) + OH^-(aq)$$

The $K_b$ of this second dissociation is $8.9 \times 10^{-16}$, so it contributes almost no hydroxide ions in solution and can be ignored here.

**a.** The pH of a 0.120 M solution of hydrazine at 25°C is 10.50. Calculate $K_b$ for the first ionization of hydrazine. Assume that the original concentration of $H_2NNH_2$ does not change.

**b.** Make the same assumption as you did in (a) and calculate the $[OH^-]$ of a 0.020 M solution.

**c.** Calculate the pH of the solution in (b).

# Problem Solving

## Redox Equations

The feature that distinguishes redox reactions from other types of reactions is that elements change oxidation state by gaining or losing electrons. Compare the equations for the following two reactions:

$$(1)\ \text{KBr}(aq) + \text{AgNO}_3(aq) \rightarrow \text{AgBr}(s) + \text{KNO}_3(aq)$$

$$(2)\ 2\text{KBr}(aq) + \text{Cl}_2(g) \rightarrow 2\text{KCl}(aq) + \text{Br}_2(l)$$

Equation 1 represents the combining of the salts potassium bromide solution and a silver nitrate solution to form a precipitate of insoluble silver bromide, leaving potassium nitrate in solution. The only change that occurs is that ions trade places, forming an insoluble compound. This reaction is a typical double-displacement reaction. It is driven by the removal of $\text{Ag}^+$ and $\text{Br}^-$ from solution in the form of a precipitate.

You will recognize that Equation 2 is a single-displacement reaction in which chlorine atoms replace bromine atoms in the salt KBr. Although it is not complex, Equation 2 differs from Equation 1 in a fundamental way. In order for chlorine to replace bromine, the uncharged atoms of elemental chlorine must change into chloride ions, each having a $1-$ charge. Also, bromide ions with a $1-$ charge must change into uncharged bromine atoms. The $\text{K}^+$ is a spectator ion that doesn't participate in the process. In fact, it could be $\text{Na}^+$, $\text{Ca}^{2+}$, $\text{Fe}^{3+}$, $\text{H}^+$, or any other stable cation.

The loss of electrons by bromine is oxidation, and the gain of electrons by chlorine is reduction.

Formation of the chloride ions and the bromine molecule involves the complete transfer of two electrons. The two chlorine atoms gain two electrons, and two bromide ions lose two electrons. Oxidation and reduction can involve the partial transfer of electrons as well as the complete transfer seen in the preceding example. The oxidation number of an atom is not synonymous with the charge on that atom, so a change in oxidation number does not require a change in actual charge. Take the example of the following half-reaction:

$$\text{H}_2\text{C}_2\text{O}_4(aq) \rightarrow 2\text{CO}_2(g) + 2\text{H}^+(aq) + 2e^-$$

Carbon changes oxidation state from $+3$ to $+4$. Carbon is oxidized even though it is not ionized.

The potassium bromide–chlorine reaction is simple, but many redox reactions are not. In this worksheet, you will practice the art of balancing redox equations and try your hand at more-complex ones.

**Problem Solving** *continued*

There are seven simple steps to balancing redox equations. You will find these steps in the General Plan for Balancing Redox Equations.

### General Plan for Balancing Redox Equations

**1** Write the unbalanced formula equation if it is not given. List formulas for any ionic substances as their individual ions, and write a total ionic equation.

**2** Assign oxidation numbers to each element. Then rewrite the equation, leaving out any ions or molecules whose elements do not change oxidation state during the reaction.

**3** Write the half-reaction for reduction. You must decide which element of the ions and molecules left after item 2 is reduced. Once you have written the half-reaction, you must balance it for charge and mass.

**4** Write the half-reaction for oxidation. You must decide which element of the ions and molecules left after item 2 is oxidized. Once you have written the half-reaction, you must balance it for charge and mass.

**5** Adjust the coefficients of the two half-reactions so that the same number of electrons are gained in reduction as are lost in oxidation.

**6** Add the two half-reactions together. Cancel out anything common to both sides of the new equation. Note that the electrons should always cancel out of the total equation.

**7** Combine ions to form the compounds shown in the original formula equation. Check to ensure that all other ions and atoms balance.

# REACTIONS IN ACIDIC SOLUTION

## Sample Problem 1

Write a balanced redox equation for the reaction of hydrochloric acid with nitric acid to produce aqueous hypochlorous acid and nitrogen monoxide.

## Solution

### ANALYZE

What is given in the problem?    **the reactants and products of a redox reaction**

What are you asked to find?    **the balanced redox reaction**

| Items | Data |
|---|---|
| Reactants | $HCl$, $HNO_3$ |
| Products | $HClO$, $NO$ |
| Solution type | acidic |
| Oxidized species | ? |
| Reduced species | ? |
| Balanced equation | ? |

### PLAN

What steps are needed to balance the redox equation?

1. Write the unbalanced formula equation followed by the ionic equation.
2. Assign oxidation numbers to each element. Delete any ion or molecule in which there is no change in oxidation state.
3. Write the half-reaction for reduction, and balance the mass and charge. $H^+$ and $H_2O$ may be added to either side of the equation to balance mass.
4. Repeat step 3 for the oxidation half-reaction.
5. Adjust the coefficients so that the number of electrons lost equals the number of electrons gained.
6. Combine the half-reactions, and cancel anything common to both sides of the equation.
7. Combine ions to change the equation back to its original form, and check the balance of everything.

### COMPUTE

1. Write the formula equation.

$$HCl(aq) + HNO_3(aq) \rightarrow HClO(aq) + NO(g)$$

Write the total ionic equation.

$$H^+ + Cl^- + H^+ + NO_3^- \rightarrow H^+ + ClO^- + NO$$

**2.** Assign oxidation numbers to each element.

$$\overset{+1}{H^+} + \overset{-1}{Cl^-} + \overset{+1}{H^+} + \overset{+5-2}{NO_3^-} \rightarrow \overset{+1}{H^+} + \overset{-2+1}{OCl^-} + \overset{+2-2}{NO}$$

Delete any ion or molecule in which there is no change in oxidation state.

$$\overset{-1}{Cl^-} + \overset{+5}{NO_3^-} \rightarrow \overset{+1}{OCl^-} + \overset{+2}{NO}$$

**3.** Write the half-reaction for reduction.

$$\overset{+5}{NO_3^-} \rightarrow \overset{+2}{NO}$$

Balance the mass by adding $H^+$ and $H_2O$.

$$4H^+ + \overset{+5}{NO_3^-} \rightarrow \overset{+2}{O} + 2H_2O$$

Balance the charge by adding electrons to the side with the higher positive charge.

$$4H^+ + \overset{+5}{NO_3^-} + 3e^- \rightarrow \overset{+2}{O} + 2H_2O$$

**4.** Write the half-reaction for oxidation.

$$\overset{-1}{Cl^-} \rightarrow \overset{+1}{OCl^-}$$

Balance the mass by adding $H^+$ and $H_2O$.

$$\overset{-1}{Cl^-} + H_2O \rightarrow \overset{+1}{OCl^-} + 2H^+$$

Balance charge by adding electrons to the side with the higher positive charge.

$$\overset{-1}{Cl^-} + H_2O \rightarrow \overset{+1}{OCl^-} + 2H^+ + 2e^-$$

**5.** Multiply by factors so that the number of electrons lost equals the number of electrons gained.

$2e^-$ are lost in oxidation; $3e^-$ are gained in reduction. Therefore, to get $6e^-$ on both sides, calculate as follows:

$$2 \times [4H^+ + NO_3^- + 3e^- \rightarrow NO + 2H_2O]$$
$$3 \times [Cl^- + H_2O \rightarrow OCl^- + 2H^+ + 2e^-]$$

**6.** Combine the half-reactions.

$$2 \times [4H^+ + NO_3^- + 3e^- \rightarrow NO + 2H_2O]$$
$$+ 3 \times [Cl^- + H_2O \rightarrow OCl^- + 2H^+ + e^-]$$
$$\overline{3Cl^- + 3H_2O + 2NO_3^- + 8H^+ + 6e^- \rightarrow 3OCl^- + 6H^+ + 6e^- + 2NO + 4H_2O}$$

Cancel out anything common to both sides of the equation.

$$3Cl^- + 3\overset{2}{\cancel{H_2O}} + 2NO_3^- + \cancel{8}H^+ + \cancel{6e^-} \rightarrow 3OCl^- + \cancel{6}H^+ + \cancel{6e^-} + 2NO + \cancel{4}H_2O$$

**7.** Combine ions to change the equation back to its original form. There must be five $H^+$ ions on the reactant side of the equation to bind with the three chloride ions and two nitrate ions, so three $H^+$ ions must be added to each side.

$$3Cl^- + 2NO_3^- + 5H^+ \rightarrow 3OCl^- + 2NO + H_2O + 3H^+$$

$$3HCl + 2HNO_3 \rightarrow 3HOCl + 2NO + H_2O$$

Check the balance.

| Problem Solving *continued*

## EVALUATE

Are the units correct?
NA

Is the number of significant figures correct?
NA

Is the answer reasonable?
Yes; the reaction has the reactants and products required and is balanced.

# Practice

Balance the following redox equations. Assume that all reactions take place in an acid environment where $H^+$ and $H_2O$ are readily available.

1. $Fe + SnCl_4 \rightarrow FeCl_3 + SnCl_2$ ans: $2Fe + 3SnCl_4 \rightarrow 2FeCl_3 + 3SnCl_2$

2. $H_2O_2 + FeSO_4 + H_2SO_4 \rightarrow Fe_2(SO_4)_3 + H_2O$
   ans: $H_2O_2 + 2FeSO_4 + H_2SO_4 \rightarrow Fe_2(SO_4)_3 + 2H_2O$

3. $CuS + HNO_3 \rightarrow Cu(NO_3)_2 + NO + S + H_2O$ ans: $3CuS + 8HNO_3 \rightarrow$
   $3Cu(NO_3)_2 + 2NO + 3S + 4H_2O$

4. $K_2Cr_2O_7 + HI \rightarrow CrI_3 + KI + I_2 + H_2O$ ans: $K_2Cr_2O_7 + 14HI \rightarrow$
   $2CrI_3 + 2KI + 3I_2 + 7H_2O$

**Problem Solving** *continued*

## REACTIONS IN BASIC SOLUTION

## Sample Problem 2

Write a balanced equation for the reaction in a basic solution of $NiO_2$ and Fe to produce $Ni(OH)_2$ and $Fe(OH)_2$.

## Solution

### ANALYZE

What is given in the problem?  **the reactants and products of a redox reaction**

What are you asked to find?  **the balanced redox reaction**

| Items | Data |
|---|---|
| Reactants | $NiO_2$, Fe |
| Products | $Ni(OH)_2$, $Fe(OH)_2$ |
| Solution type | basic |
| Oxidized species | ? |
| Reduced species | ? |
| Balanced equation | ? |

### PLAN

What steps are needed to balance the redox equation?

1. **Write the formula equation followed by the ionic equation.**
2. **Assign oxidation numbers to each element. Delete any ion or molecule in which there is no change in oxidation state.**
3. **Write the half-reaction for reduction, and balance the mass and charge. $OH^-$ and $H_2O$ may be added to either side.**
4. **Repeat step 3 for the oxidation half-reaction.**
5. **Adjust the coefficients so that the number of electrons lost equals the number of electrons gained.**
6. **Combine the half-reactions, and cancel anything common to both sides of the equation.**
7. **Combine ions to change the equation back to its original form, and check the balance of everything.**

### COMPUTE

1. Write the formula equation.

$$NiO_2 + Fe \rightarrow Ni(OH)_2 + Fe(OH)_2$$

Write the total ionic equation.

$$NiO_2 + Fe \rightarrow Ni^{2+} + 2OH^- + Fe^{2+} + 2OH^-$$

**Problem Solving** *continued*

**2.** Assign oxidation numbers to each element.

$$\overset{+4\,-2}{NiO_2} + \overset{0}{Fe} \rightarrow \overset{+2}{Ni}^{2+} + 2\overset{-2\,+1}{OH}^- + \overset{+2}{Fe}^{2+} + 2\overset{-2\,+1}{OH}^-$$

Delete any ions or molecules in which there is no change in oxidation state.

$$\overset{+4}{NiO_2} + \overset{0}{Fe} \rightarrow \overset{+2}{Ni}^{2+} + \overset{+2}{Fe}^{2+}$$

**3.** Write the half-reaction for reduction.

$$\overset{+4}{NiO_2} \rightarrow \overset{+2}{Ni}^{2+}$$

Balance the mass by adding $OH^-$ and $H_2O$.

$$\overset{+4}{NiO_2} + 2H_2O \rightarrow \overset{+2}{Ni}^{2+} + 4OH^-$$

Balance the charge by adding electrons to the side with the higher positive charge.

$$\overset{+4}{NiO_2} + 2H_2O + 2e^- \rightarrow \overset{+2}{Ni}^{2+} + 4OH^-$$

**4.** Write the half-reaction for oxidation.

$$\overset{0}{Fe} \rightarrow \overset{+2}{Fe}^{2+}$$

The mass is already balanced.
Balance the charge by adding electrons to the side with the higher positive charge.

$$\overset{0}{Fe} \rightarrow \overset{+2}{Fe}^{2+} + 2e^-$$

**5.** The numbers of electrons lost and gained are already the same.

**6.** Combine the half-reactions.

$$\overset{+4}{NiO_2} + 2H_2O + 2e^- \rightarrow \overset{+2}{Ni}^{2+} + 4OH^-$$
$$\underline{+ \overset{0}{Fe} \rightarrow \overset{+2}{Fe}^{2+} + 2e^-}$$
$$NiO_2 + 2H_2O + Fe \rightarrow Ni^{2+} + Fe^{2+} + 4OH^-$$

**7.** Combine ions to change the equation back to its original form. The four $OH^-$ ions combine with the nickel and iron to make nickel(II) hydroxide and iron(II) hydroxide.

$$NiO_2 + 2H_2O + Fe \rightarrow Ni(OH)_2 + Fe(OH)_2$$

Check the balance.

**EVALUATE**

Are the units correct?
**NA**

Is the number of significant figures correct?
**NA**

Is the answer reasonable?
**Yes; the reaction has the reactants and products required and is balanced.**

| **Problem Solving** *continued* |

## Practice

**Balance the following redox equations. Assume that all reactions take place in a basic environment where OH$^-$ and H$_2$O are readily available.**

**1.** $CO_2 + NH_2OH \rightarrow CO + N_2 + H_2O$ **ans: $CO_2 + 2NH_2OH \rightarrow CO + N_2 + 3H_2O$**

**2.** $Bi(OH)_3 + K_2SnO_2 \rightarrow Bi + K_2SnO_3$ (Both of the potassium-tin-oxygen compounds dissociate into potassium ions and tin-oxygen ions.)
   **ans: $2Bi(OH)_3 + 3K_2SnO_2 \rightarrow 2Bi + 3K_2SnO_3 + 3H_2O$**

**▌Problem Solving** *continued*

## Additional Problems

**Balance each of the following redox equations. Unless stated otherwise, assume that the reaction occurs in acidic solution.**

1. $Mg + N_2 \rightarrow Mg_3N_2$

2. $SO_2 + Br_2 + H_2O \rightarrow HBr + H_2SO_4$

3. $H_2S + Cl_2 \rightarrow S + HCl$

4. $PbO_2 + HBr \rightarrow PbBr_2 + Br_2 + H_2O$

5. $S + HNO_3 \rightarrow NO_2 + H_2SO_4 + H_2O$

6. $NaIO_3 + N_2H_4 + HCl \rightarrow N_2 + NaICl_2 + H_2O$ ($N_2H_4$ is hydrazine; do not separate it into ions.)

7. $MnO_2 + H_2O_2 + HCl \rightarrow MnCl_2 + O_2 + H_2O$

8. $AsH_3 + NaClO_3 \rightarrow H_3AsO_4 + NaCl$ ($AsH_3$ is arsine, the arsenic analogue of ammonia, $NH_3$.)

9. $K_2Cr_2O_7 + H_2C_2O_4 + HCl \rightarrow CrCl_3 + CO_2 + KCl + H_2O$ ($H_2C_2O_4$ is oxalic acid; it can be treated as $2H^+ + C_2O_4^{2-}$.)

10. $Hg(NO_3)_2 \xrightarrow{\triangle} HgO + NO_2 + O_2$ (The reaction is not in solution.)

11. $HAuCl_4 + N_2H_4 \rightarrow Au + N_2 + HCl$ ($HAuCl_4$ can be considered as $H^+ + AuCl_4^-$.)

12. $Sb_2(SO_4)_3 + KMnO_4 + H_2O \rightarrow H_3SbO_4 + K_2SO_4 + MnSO_4 + H_2SO_4$

13. $Mn(NO_3)_2 + NaBiO_3 + HNO_3 \rightarrow Bi(NO_3)_2 + HMnO_4 + NaNO_3 + H_2O$

14. $H_3AsO_4 + Zn + HCl \rightarrow AsH_3 + ZnCl_2 + H_2O$

15. $KClO_3 + HCl \rightarrow Cl_2 + H_2O + KCl$

16. The same reactants as in Item 15 can combine in the following way when more $KClO_3$ is present. Balance the equation.
$$KClO_3 + HCl \rightarrow Cl_2 + ClO_2 + H_2O + KCl$$

17. $MnCl_3 + H_2O \rightarrow MnCl_2 + MnO_2 + HCl$

18. $NaOH + H_2O + Al \rightarrow NaAl(OH)_4 + H_2$ in basic solution

19. $Br_2 + Ca(OH)_2 \rightarrow CaBr_2 + Ca(BrO_3)_2 + H_2O$ in basic solution

20. $N_2O + NaClO + NaOH \rightarrow NaCl + NaNO_2 + H_2O$ in basic solution

21. Balance the following reaction, which can be used to prepare bromine in the laboratory:
$$HBr + MnO_2 \rightarrow MnBr_2 + H_2O + Br_2$$

22. The following reaction occurs when gold is dissolved in *aqua regia*. Balance the equation.
$$Au + HCl + HNO_3 \rightarrow HAuCl_4 + NO + H_2O$$

# Problem Solving

## Electrochemistry

The potential in volts has been measured for many different reduction half-reactions. The potential value is measured against the standard hydrogen electrode, which is assigned a value of zero. For consistency, these half-reactions are always written in the direction of the reduction. A half-reaction that has a positive reduction potential proceeds in the direction of the reduction when it is coupled with a hydrogen electrode. A reaction that has a negative reduction potential proceeds in the oxidation direction when it is coupled with a hydrogen electrode. **Table 1** gives some common standard reduction potentials.

**TABLE 1**

| Reduction half-reaction | Standard electrode potential, $E^0$ (in volts) | Reduction half-reaction | Standard electrode potential, $E^0$ (in volts) |
|---|---|---|---|
| $MnO_4^- + 8H^+ + 5e^- \rightleftarrows Mn^{2+} + 4H_2O$ | +1.50 | $Fe^{3+} + 3e^- \rightleftarrows Fe$ | −0.04 |
| $Au^{3+} + 3e^- \rightleftarrows Au$ | +1.50 | $Pb^{2+} + 2e^- \rightleftarrows Pb$ | −0.13 |
| $Cl_2 + 2e^- \rightleftarrows 2Cl^-$ | +1.36 | $Sn^{2+} + 2e^- \rightleftarrows Sn$ | −0.14 |
| $Cr_2O_7^{2-} + 14H^+ + 6e^- \rightleftarrows 2Cr^{3+} + 7H_2O$ | +1.23 | $Ni^{2+} + 2e^- \rightleftarrows Ni$ | −0.26 |
| $MnO_2 + 4H^+ + 2e^- \rightleftarrows Mn^{2+} + 2H_2O$ | +1.22 | $Cd^{2+} + 2e^- \rightleftarrows Cd$ | −0.40 |
| $Br_2 + 2e^- \rightleftarrows 2Br^-$ | +1.07 | $Fe^{2+} + 2e^- \rightleftarrows Fe$ | −0.45 |
| $Hg^{2+} + 2e^- \rightleftarrows Hg$ | +0.85 | $S + 2e^- \rightleftarrows S^{2-}$ | −0.48 |
| $Ag^+ + e^- \rightleftarrows Ag$ | +0.80 | $Zn^{2+} + 2e^- \rightleftarrows Zn$ | −0.76 |
| $Hg_2^{2+} + 2e^- \rightleftarrows 2Hg$ | +0.80 | $Al^{3+} + 3e^- \rightleftarrows Al$ | −1.66 |
| $Fe^{3+} + e^- \rightleftarrows Fe^{2+}$ | +0.77 | $Mg^{2+} + 2e^- \rightleftarrows Mg$ | −2.37 |
| $MnO_4^- + e^- \rightleftarrows MnO_4^{2-}$ | +0.56 | $Na^+ + e^- \rightleftarrows Na$ | −2.71 |
| $I_2 + 2e^- \rightleftarrows 2I^-$ | +0.54 | $Ca^{2+} + 2e^- \rightleftarrows Ca$ | −2.87 |
| $Cu^{2+} + 2e^- \rightleftarrows Cu$ | +0.34 | $Ba^{2+} + 2e^- \rightleftarrows Ba$ | −2.91 |
| $S + 2H^+(aq) + 2e^- \rightleftarrows H_2S(aq)$ | +0.14 | $K^+ + e^- \rightleftarrows K$ | −2.93 |
| $2H^+(aq) + 2e^- \rightleftarrows H_2$ | +0.00 | $Li^+ + e^- \rightleftarrows Li$ | −3.04 |

**Problem Solving** *continued*

You can use reduction potentials to predict the direction in which any redox reaction will be spontaneous. A spontaneous reaction occurs by itself, without outside influence. The redox reaction will proceed in the direction for which the difference between the two half-reaction potentials is positive. This direction is the same as the direction of the more positive half-reaction.

**General Plan for Solving Electrochemical Problems**

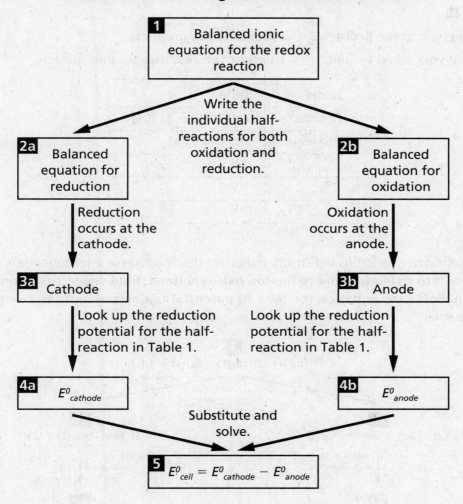

## Sample Problem 1

**Calculate the cell potential to determine whether the following reaction is spontaneous in the direction indicated.**

$$Cd^{2+}(aq) + 2I^-(aq) \rightarrow Cd(s) + I_2(s)$$

## Solution

### ANALYZE

What is given in the problem?　　**reactants and products**

What are you asked to find?　　**whether the reaction is spontaneous**

| Items | Data |
|---|---|
| Reactants | $Cd^{2+}(aq) + 2I^-(aq)$ |
| Products | $Cd(s) + I_2(s)$ |
| $E^O_{cathode}$ | ? V |
| $E^O_{anode}$ | ? V |
| $E^O_{cell}$ | ? V |

### PLAN

What steps are needed to determine whether the reaction is spontaneous?
**Separate into oxidation and reduction half-reactions. Find reduction potentials for each. Solve the equation for the cell potential to determine if the reaction is spontaneous.**

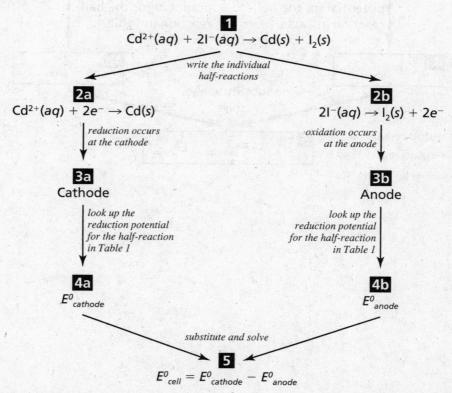

**1**
$$Cd^{2+}(aq) + 2I^-(aq) \rightarrow Cd(s) + I_2(s)$$

*write the individual half-reactions*

**2a**
$$Cd^{2+}(aq) + 2e^- \rightarrow Cd(s)$$

*reduction occurs at the cathode*

**3a**
Cathode

*look up the reduction potential for the half-reaction in Table 1*

**4a**
$E^O_{cathode}$

**2b**
$$2I^-(aq) \rightarrow I_2(s) + 2e^-$$

*oxidation occurs at the anode*

**3b**
Anode

*look up the reduction potential for the half-reaction in Table 1*

**4b**
$E^O_{anode}$

*substitute and solve*

**5**
$$E^O_{cell} = E^O_{cathode} - E^O_{anode}$$

**| Problem Solving** *continued*

Write the given equation.

$$Cd^{2+}(aq) + 2I^-(aq) \rightarrow Cd(s) + I_2(s)$$

The oxidation number of cadmium decreases; it is reduced.

$$Cd^{2+}(aq) + 2e^- \rightarrow Cd(s)$$

The oxidation number of iodine increases; it is oxidized.

$$2I^-(aq) \rightarrow I_2(s) + 2e^-$$

Cadmium is the cathode, and iodine is the anode.

$$E^0_{cathode} = \overset{from\ Table\ 1}{-0.40\ V}$$

$$E^0_{anode} = \overset{from\ Table\ 1}{+0.54\ V}$$

$$E^0_{cell} = \overset{given\ above}{E^0_{cathode}} - \overset{given\ above}{E^0_{anode}}$$

Determine spontaneity. If the cell potential is positive, the reaction is sponta-neous as written. If the cell potential is negative, the reaction is not spontaneous as written, but the reverse reaction is spontaneous.

**COMPUTE**

$$E^0_{cell} = -0.40\ V - 0.54\ V = -0.94\ V$$

The reaction potential is negative. Therefore, the reaction is not spontaneous. The reverse reaction would have a positive potential and would, therefore, be spontaneous.

$$Cd^{2+}(aq) + 2I^-(aq) \rightarrow Cd(s) + I_2(s)\ not\ spontaneous$$

$$Cd(s) + I_2(s) \rightarrow Cd^{2+}(aq) + 2I^-(aq)\ spontaneous$$

**EVALUATE**

Are the units correct?
**Yes; cell potentials are in volts.**

Is the number of significant figures correct?
**Yes; the number of significant figures is correct because the half-cell potentials have two significant figures.**

Is the answer reasonable?
**Yes; the reduction potential for the half-reaction involving iodine was more positive than the potential for the reaction involving cadmium, which means that $I_2$ has a greater attraction for electrons than $Cd^{2+}$. Therefore, $I_2$ is more likely to be reduced than $Cd^{2+}$. The reverse reaction is favored.**

**▌Problem Solving** *continued*

## Practice

**Use the reduction potentials in Table 1 to determine whether the following reactions are spontaneous as written. Report the $E^0_{cell}$ for the reactions.**

**1.** $Cu^{2+} + Fe \rightarrow Fe^{2+} + Cu$ **ans: +0.79 V; spontaneous**

**2.** $Pb^{2+} + Fe^{2+} \rightarrow Fe^{3+} + Pb$ **ans: −0.90 V; nonspontaneous**

**3.** $Mn^{2+} + 4H_2O + Sn^{2+} \rightarrow MnO_4^- + 8H^+ + Sn$ **ans: −1.64 V; nonspontaneous**

**4.** $MnO_4^{2-} + Cl_2 \rightarrow MnO_4^- + 2Cl^-$ **ans: +0.80 V; spontaneous**

**5.** $Hg_2^{2+} + 2MnO_4^{2-} \rightarrow 2Hg + 2MnO_4^{-}$ **ans: +0.24 V; spontaneous**

**6.** $2Li^+ + Pb \rightarrow 2Li + Pb^{2+}$ **ans: −2.91 V; nonspontaneous**

**7.** $Br_2 + 2Cl^- \rightarrow 2Br^- + Cl_2$ **ans: −0.29 V; nonspontaneous**

**8.** $S + 2I^- \rightarrow S^{2-} + I_2$ **ans: −1.02 V; nonspontaneous**

## Sample Problem 2

A cell is constructed in which the following two half-reactions can occur in either direction.

$$Zn^{2+} + 2e^- \rightleftarrows Zn$$

$$Br_2 + 2e^- \rightleftarrows 2Br^-$$

Write the full ionic equation for the cell in the spontaneous direction, identify the reactions occurring at the anode and cathode, and determine the cell's voltage.

## Solution

### ANALYZE

What is given in the problem?    the reversible half-reactions of the cell

What are you asked to find?    the equation in the spontaneous direction; the voltage of the cell

| Items | Data |
|---|---|
| Half-reaction 1 | $Zn^{2+} + 2e^- \rightleftarrows Zn$ |
| Half-reaction 2 | $Br_2 + 2e^- \rightleftarrows 2Br$ |
| Reduction potential of 1 | $-0.76$ V |
| Reduction potential of 2 | $+1.07$ V |
| Full ionic reaction | ? |
| Cell voltage | ? |

### PLAN

What steps are needed to determine the spontaneous reaction of the cell and the cell voltage?

Determine which half-reaction has the more positive reduction potential. This will be the reduction half-reaction; it occurs at the cathode. Reverse the other half-reaction so that it becomes an oxidation half-reaction; it occurs at the anode. Adjust the half-reactions so that the same number of electrons are lost as are gained. Add the reactions together. Compute the cell voltage by the formula $E^0_{cell} = E^0_{cathode} - E^0_{anode}$, using the reduction potentials for the reaction at each electrode.

**Problem Solving** *continued*

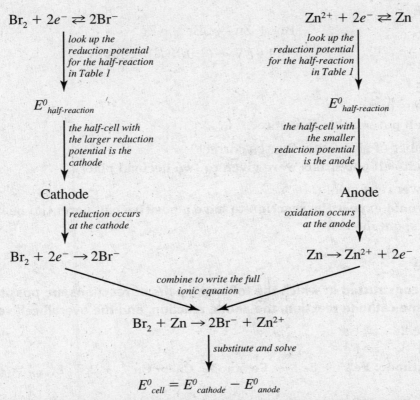

$$\text{Br}_2 + 2e^- \rightleftharpoons 2\text{Br}^-$$

*look up the
reduction potential
for the half-reaction
in Table 1*

$E^0_{\text{half-reaction}}$

*the half-cell with
the larger reduction
potential is the
cathode*

Cathode

*reduction occurs
at the cathode*

$$\text{Br}_2 + 2e^- \rightarrow 2\text{Br}^-$$

$$\text{Zn}^{2+} + 2e^- \rightleftharpoons \text{Zn}$$

*look up the
reduction potential
for the half-reaction
in Table 1*

$E^0_{\text{half-reaction}}$

*the half-cell with
the smaller
reduction potential
is the anode*

Anode

*oxidation occurs
at the anode*

$$\text{Zn} \rightarrow \text{Zn}^{2+} + 2e^-$$

*combine to write the full
ionic equation*

$$\text{Br}_2 + \text{Zn} \rightarrow 2\text{Br}^- + \text{Zn}^{2+}$$

*substitute and solve*

$$E^0_{\text{cell}} = E^0_{\text{cathode}} - E^0_{\text{anode}}$$

First, look up the reduction potentials for the two half-reactions in Table 1.

$$E^0_{Br_2} = \overset{\text{from Table 1}}{+1.07 \text{ V}}$$

$$E^0_{Zn} = \overset{\text{from Table 1}}{-0.76 \text{ V}}$$

$\text{Br}_2$ has the larger reduction potential; therefore, it is the cathode. Zn has the smaller reduction potential; therefore, it is the anode.

The cathode half-reaction is

$$\text{Br}_2 + 2e^- \rightarrow 2\text{Br}$$

The anode half-reaction is

$$\text{Zn} \rightarrow \text{Zn}^{2+} + 2e^-$$

The full-cell equation is

$$\text{Br}_2 + 2e^- \rightarrow 2\text{Br}^-$$
$$+ \text{Zn} \rightarrow \text{Zn}^{2+} + 2e^-$$
$$\overline{\text{Br}_2 + \text{Zn} \rightarrow 2\text{Br}^- + \text{Zn}^{2+}}$$

Substitute the reduction potentials for the anode and cathode into the cell potential equation, and solve the equation.

$$E^0_{\text{cell}} = E^0_{\text{cathode}}{}^{E^0_{Br_2}} - E^0_{\text{anode}}{}^{E^0_{Zn}}$$

**COMPUTE**

$$Br_2 + Zn \rightarrow 2Br^- + Zn^{2+}$$
$$E^O{}_{cell} = 1.07\ V - (-0.76\ V) = 1.83\ V$$

**EVALUATE**

Are the units correct?

**Yes; the cell potential is in volts.**

Is the number of significant figures correct?

**Yes; the half-cell potentials were given to two decimal places.**

Is the answer reasonable?

**Yes; you would expect the reaction to have a positive cell potential because it should be spontaneous.**

# Practice

**If a cell is constructed in which the following pairs of reactions are possible, what would be the cathode reaction, the anode reaction, and the overall cell voltage?**

**1.** $Ca^{2+} + 2e^- \rightleftarrows Ca$
$Fe^{3+} + 3e^- \rightleftarrows Fe$
**ans: cathode: $Fe^{3+} + 3e^- \rightarrow Fe$, anode: $Ca \rightarrow Ca^{2+} + 2e^-$, $E^O{}_{cell} = +2.83\ V$**

**2.** $Ag^+ + e^- \rightleftarrows Ag$
$S + 2H^+ + 2e^- \rightleftarrows H_2S$
**ans: cathode: $Ag^+ + e^- \rightarrow Ag$, anode: $H_2S \rightarrow S + 2H^+ + 2e^-$, $E^O{}_{cell} = +0.66\ V$**

**3.** $Fe^{3+} + e^- \rightleftarrows Fe^{2+}$
$Sn^{2+} + 2e^- \rightleftarrows Sn$
**ans: cathode: $Fe^{3+} + e^- \rightarrow Fe^{2+}$, anode: $Sn \rightarrow Sn^{2+} + 2e^-$, $E^O{}_{cell} = +0.91\ V$**

**4.** $Cu^{2+} + 2e^- \rightleftarrows Cu$
$Au^{3+} + 3e^- \rightleftarrows Au$
**ans: cathode: $Au^{3+} + 3e^- \rightarrow Au$, anode: $Cu \rightarrow Cu^{2+} + 2e^-$, $E^O{}_{cell} = +1.16\ V$**

**❚ Problem Solving** *continued*

# Additional Problems

**Use reduction potentials to determine whether the reactions in the following 10 problems are spontaneous.**

**1.** $Ba + Sn^{2+} \rightarrow Ba^{2+} + Sn$

**2.** $Ni + Hg^{2+} \rightarrow Ni^{2+} + Hg$

**3.** $2Cr^{3+} + 7H_2O + 6Fe^{3+} \rightarrow Cr_2O_7^{2-} + 14H^+ + 6Fe^{2+}$

**4.** $Cl_2 + Sn \rightarrow 2Cl^- + Sn^{2+}$

**5.** $Al + 3Ag^+ \rightarrow Al^{3+} + 3Ag$

**6.** $Hg_2^{2+} + S^{2-} \rightarrow 2Hg + S$

**7.** $Ba + 2Ag^+ \rightarrow Ba^{2+} + 2Ag$

**8.** $2I^- + Ca^{2+} \rightarrow I_2 + Ca$

**9.** $Zn + 2MnO_4^- \rightarrow Zn^{2+} + 2MnO_4^{2-}$

**10.** $2Cr^{3+} + 3Mg^{2+} + 7H_2O \rightarrow Cr_2O_7^{2-} + 14H^+ + 3Mg$

**In the following problems, you are given a pair of reduction half-reactions. If a cell were constructed in which the pairs of half-reactions were possible, what would be the balanced equation for the overall cell reaction that would occur? Write the half-reactions that occur at the cathode and anode, and calculate the cell voltage.**

**11.** $Cl_2 + 2e^- \rightleftarrows 2Cl^-$
$\quad\ Ni^{2+} + 2e^- \rightleftarrows Ni$

**12.** $Fe^{3+} + 3e^- \rightleftarrows Fe$
$\quad\ Hg^{2+} + 2e^- \rightleftarrows Hg$

**13.** $MnO_4^- + e^- \rightleftarrows MnO_4^{2-}$
$\quad\ Al^{3+} + 3e^- \rightleftarrows Al$

**14.** $MnO_4^- + 8H^+ + 5e^- \rightleftarrows Mn^{2+} + 4H_2O$
$\quad\ S + 2H^+ + 2e^- \rightleftarrows H_2S$

**15.** $Ca^{2+} + 2e^- \rightleftarrows Ca$
$\quad\ Li^+ + e^- \rightleftarrows Li$

**16.** $Br_2 + 2e^- \rightleftarrows 2Br^-$
$\quad\ MnO_4^- + 8H^+ + 5e^- \rightleftarrows Mn^{2+} + 4H_2O$

**17.** $Sn^{2+} + 2e^- \rightleftarrows Sn$
$\quad\ Fe^{3+} + e^- \rightleftarrows Fe^{2+}$

**18.** $Zn^{2+} + 2e^- \rightleftarrows Zn$
$\quad\ Cr_2O_7^{2-} + 14H^+ + 6e^- \rightleftarrows 2Cr^{3+} + 7H_2O$

**19.** $Ba^{2+} + 2e^- \rightleftarrows Ba$
$\quad\ Ca^{2+} + 2e^- \rightleftarrows Ca$

**20.** $Hg_2^{2+} + 2e^- \rightleftarrows 2Hg$
$\quad\ Cd^{2+} + 2e^- \rightleftarrows Cd$

# Answer Key

## The Science of Chemistry

### CONVERSIONS

1. **a.** 12 750 km
   **b.** 2.77 m
   **c.** 3.056 hectares
   **d.** 0.008 19 $m^2$
   **e.** 300 Mm
2. **a.** 620 m
   **b.** 3 875 000 mg
   **c.** 3.6 $\mu$L
   **d.** 342 kg
   **e.** 68 710 L
3. **a.** 0.000 856 kg
   **b.** 0.001 21 kg
   **c.** 6.598 $cm^3$
   **d.** 0.0806 mm
   **e.** 0.010 74 L
4. **a.** 7930 $cm^3$
   **b.** 590 cm
   **c.** 4.19 $dm^3$
   **d.** 74 800 $cm^2$
   **e.** 197 L
5. 1 L
6. 370 $\mu$L
7. 6 L
8. 0.876 L of water per year; 876 kg of water
9. **a.** 1674 km/h
   **b.** 40 176 km/day
10. 7.8 kg of sodium hydroxide
11. 45 m of plastic tubing
12. **a.** 13.2 mL/day
    **b.** 150 kg/min
    **c.** 62 $cm^3$/min
    **d.** 1.7 m/s
13. **a.** 2.97 $g/cm^3$
    **b.** 0.041 28 $kg/cm^2$
    **c.** 5.27 $kg/dm^3$
    **d.** 0.006 91 $mg/mm^3$
14. **a.** 750 mL
    **b.** 5.56 kg
15. 1250 kg
16. **a.** 0.0028$m^3$
    **b.** 1.05 $g/cm^3$
    **c.** 0.056 $m^2$
17. **a.** 0.04 mL per drop
    **b.** 1.48 mL
    **c.** 17 000 drops

18. **a.** 0.5047 kg; 504.7 g
    **b.** 0.0092 kg; 9.2 g
    **c.** 0.000 122 kg; 0.122 g
    **d.** 0.071 95 kg; 71.95 g
19. **a.** 0.582 L; 582 mL
    **b.** 2.5 L; 2500 mL
    **c.** 1.18 L; 1180 mL
    **d.** 0.0329 L; 32.9 mL
20. **a.** 1370 g/L; 1370 $kg/m^3$
    **b.** 692 g/L; 692 $kg/m^3$
    **c.** 5200 g/L; 5200 $kg/m^3$
    **d.** 38 g/L; 38 $kg/m^3$
    **e.** 5790 g/L; 5790 $kg/m^3$
    **f.** 0.0011 g/L; 0.0011 $kg/m^3$
21. **a.** 360 g/min
    **b.** 518.4 kg/day
    **c.** 6 mg/ms
22. 27.8 m/s
23. 4732 kcal/h
24. 620 kg
25. $\dfrac{3.9 \text{ mL}}{h} \times \dfrac{1 \text{ L}}{1000 \text{ mL}} \times \dfrac{24 \text{ h}}{1 \text{ day}} \times$
    $\dfrac{365 \text{ days}}{1 \text{ year}} = 34.164$ L/year

26. 40 doses

## Matter and Energy

### SIGNIFICANT FIGURES

1. **a.** 3
   **b.** 4
   **c.** 3
   **d.** 2
   **e.** 2
   **f.** 1
   **g.** 3
   **h.** 4
   **i.** 5
2. **a.** 5 490 000 m
   **b.** 0.013 479 3 mL
   **c.** 31 950 $cm^2$
   **d.** 192.67 $m^2$
   **e.** 790 cm
   **f.** 389 278 000 J
   **g.** 225 834.8 $cm^3$
3. **a.** 49 000 $cm^2$
   **b.** 3.1 kg/L
   **c.** 12.3 L/sec
   **d.** 170 000 $cm^3$

**e.** 41 m$^3$
**f.** 3.129 g/cm$^3$
**4. a.** 90.2 J
  **b.** 0.0006 m
  **c.** 900 g
  **d.** 31.1 kPa
  **e.** 278 dL
  **f.** 1790 kg
**5. a.** 307 cm$^2$
  **b.** 30 700 mm$^2$
  **c.** 0.0307 m$^2$
**6. a.** 1800 cm$^3$
  **b.** 0.0018 m$^3$
  **c.** 1 800 000 mm$^3$
**7. a.** 1300 kg/m$^3$
  **b.** 1.3 g/mL
  **c.** 1.3 kg/dm$^3$
**8. a.** 130 mm$^3$
  **b.** 430 cm$^3$
  **c.** 5.0 m
  **d.** 4000 m$^3$
**9.** 26 000 000 m$^3$
**10. a.** 13.38 g
  **b.** 100. mg
  **c.** 0.015 L
  **d.** 315 cm$^2$
  **e.** 14.47 kg
  **f.** 353 mL
**11.** 1.09 kg/L
**12.** 0.43 g/m; 2.3 m
**13.** 2000 m$^2$
**14.** 26 300 kJ/min; 439 kJ/s
**15. a.** 15.8 m$^3$
  **b.** 9800 L/min
  **c.** 590 m$^3$/h
**16. a.** 7.5 kg·m$^2$
  **b.** 67.22 cm
  **c.** 2.4 kg·m$^2$/s$^2$
  **d.** 19.9 m$^2$
  **e.** 970 000 m/h
  **f.** 139 cm$^2$

## SCIENTIFIC NOTATION
**1. a.** $1.58 \times 10^5$ km
  **b.** $9.782 \times 10^{-6}$ L
  **c.** $8.371 \times 10^8$ cm$^3$
  **d.** $6.5 \times 10^9$ mm$^2$
  **e.** $5.93 \times 10^{-3}$ g
  **f.** $6.13 \times 10^{-9}$ m
  **g.** $1.2552 \times 10^7$ J
  **h.** $8.004 \times 10^{-6}$ g/L
  **i.** $1.0995 \times 10^{-2}$ kg
  **j.** $1.05 \times 10^9$ Hz

**2. a.** $9.49 \times 10^3$ kg
  **b.** $7.1 \times 10^{-2}$ mg
  **c.** $9.8 \times 10^3$ m$^3$
  **d.** $1.56 \times 10^{-7}$ m
  **e.** $3.18 \times 10^6$ J
  **f.** $9.63 \times 10^{27}$ molecules
  **g.** $7.47 \times 10^6$ cm
**3. a.** $6.53 \times 10^2$ L/s
  **b.** $1.83 \times 10^7$ mm$^3$
  **c.** $2.51 \times 10^4$ kg/m$^2$
  **d.** $4.23 \times 10^1$ km/s
  **e.** $3.22 \times 10^6$ m$^3$
  **f.** $8.68 \times 10^6$ J/s
**4.** $3.6 \times 10^6$ J
**5.** $1.12 \times 10^7$ Pa
**6.** $1.5 \times 10^6$ mm
**7. a.** $3 \times 10^5$ km/s
  **b.** $1 \times 10^{12}$ m/h
  **c.** $3 \times 10^4$ cm
**8. a.** $7.75 \times 10^{22}$ molecules
  **b.** $1.59 \times 10^{27}$ molecules
  **c.** $6.41 \times 10^{17}$ molecules
**9. a.** $2.2 \times 10^{-5}$ mm$^2$/transistor
  **b.** $4.3 \times 10^9$ transistors
**10.** $5.01 \times 10^{-8}$ g/μL
**11.** $4.79 \times 10^7$ cesium atoms
**12.** $1.2 \times 10^{20}$ g/m$^3$; $1.2 \times 10^{14}$ kg
**13.** $1.9 \times 10^7$ pits
**14. a.** $2.69 \times 10^{18}$ molecules of oxygen
  **b.** $2.69 \times 10^{22}$ molecules of oxygen
  **c.** $3.72 \times 10^{-20}$ mL/molecule
**15. a.** $7.9 \times 10^8$ kg/person
  **b.** $7.9 \times 10^5$ metric ton/person
  **c.** $5.4 \times 10^8$ kg/person
**16.** $3.329 \times 10^5$ Earths
**17. a.** $4.8 \times 10^{-1}$ km$^3$
  **b.** $4.8 \times 10^8$ m$^3$
  **c.** 32 years
**18.** $1.0 \times 10^7$ J/day

# The Mole and Chemical Composition

## FOUR STEPS FOR SOLVING QUANTITATIVE PROBLEMS
**1.** 0.026 mm
**2.** 3.21 L
**3.** 0.80 g/cm$^3$
**4.** 21.4 g/cm$^3$
**5.** 30 boxes
**6. a.** 1.73 L
    0.120 m × 0.120 m × 0.120 m
  **b.** 9.2 g; 5.0 cm$^3$

   **c.** 60.4 kg; $1.88 \times 10^4$ dm$^3$

   **d.** 0.94 g/cm$^3$; $5.3 \times 10^{-4}$ m$^3$

   **e.** $2.5 \times 10^3$ kg; $2.7 \times 10^6$ cm$^3$

 **7.** 2.8 g/cm$^3$

 **8. a.** 0.72 $\mu$m

   **b.** $2.5 \times 10^3$ atoms

 **9.** 1300 L/min

**10.** $1.3 \times 10^6$ cal/h

**11.** 5.44 g/ cm$^3$

**12.** $2.24 \times 10^4$ cm$^3$

**13.** 32 000 uses

**14.** 2500 L

**15.** 9.5 L/min

## MOLE CONCEPT

 **1. a.** $3.7 \times 10^{-4}$ mol Pd

   **b.** 150 mol Fe

   **c.** 0.040 mol Ta

   **d.** $5.38 \times 10^{-5}$ mol Sb

   **e.** 41.1 mol Ba

   **f.** $3.51 \times 10^{-8}$ mol Mo

 **2. a.** 52.10 g Cr

   **b.** $1.5 \times 10^4$ g or 15 kg Al

   **c.** $8.23 \times 10^{-7}$ g Ne

   **d.** $3 \times 10^2$ g or 0.3 kg Ti

   **e.** 1.1 g Xe

   **f.** $2.28 \times 10^5$ g or 228 kg Li

 **3. a.** $1.02 \times 10^{25}$ atoms Ge

   **b.** $3.700 \times 10^{23}$ atoms Cu

   **c.** $1.82 \times 10^{24}$ atoms Sn

   **d.** $1.2 \times 10^{30}$ atoms C

   **e.** $1.1 \times 10^{21}$ atoms Zr

   **f.** $1.943 \times 10^{14}$ atoms K

 **4. a.** 10.00 mol Co

   **b.** 0.176 mol W

   **c.** $4.995 \times 10^{-5}$ mol Ag

   **d.** $1.6 \times 10^{-15}$ mol Pu

   **e.** $7.66 \times 10^{-7}$ mol Rn

   **f.** $1 \times 10^{-11}$ mol Ce

 **5. a.** $2.5 \times 10^{19}$ atoms Au

   **b.** $5.10 \times 10^{24}$ atoms Mo

   **c.** $4.96 \times 10^{20}$ atoms Am

   **d.** $3.011 \times 10^{26}$ atoms Ne

   **e.** $2.03 \times 10^{18}$ atoms Bi

   **f.** $9.4 \times 10^{16}$ atoms U

 **6. a.** 117 g Rb

   **b.** 223 g Mn

   **c.** $2.11 \times 10^5$ g Te

   **d.** $2.6 \times 10^{-3}$ g Rh

   **e.** $3.31 \times 10^{-8}$ g Ra

   **f.** $8.71 \times 10^{-5}$ g Hf

 **7. a.** 0.749 mol CH$_3$COOH

   **b.** 0.0213 mol Pb(NO$_3$)$_2$

   **c.** $3 \times 10^4$ mol Fe$_2$O$_3$

   **d.** $2.66 \times 10^{-4}$ mol C$_2$H$_5$NH$_2$

   **e.** $1.13 \times 10^{-5}$ mol C$_{17}$H$_{35}$COOH

   **f.** 378 mol (NH$_4$)$_2$SO$_4$

 **8. a.** 764 g SeOBr$_2$

   **b.** $4.88 \times 10^4$ g CaCO$_3$

   **c.** 2.7 g C$_{20}$H$_{28}$O$_2$

   **d.** $9.74 \times 10^{-6}$ g C$_{10}$H$_{14}$N$_2$

   **e.** 529 g Sr(NO$_3$)$_2$

   **f.** $1.23 \times 10^{-3}$ g UF$_6$

 **9. a.** $2.57 \times 10^{24}$ formula units WO$_3$

   **b.** $1.81 \times 10^{21}$ formula units Sr(NO$_3$)$_2$

   **c.** $4.37 \times 10^{25}$ molecules C$_6$H$_5$CH$_3$

   **d.** $3.08 \times 10^{17}$ molecules C$_{29}$H$_{50}$O$_2$

   **e.** $9.0 \times 10^{26}$ molecules N$_2$H$_4$

   **f.** $5.96 \times 10^{23}$ molecules C$_6$H$_5$NO$_2$

**10. a.** $1.14 \times 10^{24}$ formula units FePO$_4$

   **b.** $6.4 \times 10^{19}$ molecules C$_5$H$_5$N

   **c.** $6.9 \times 10^{20}$ molecules (CH$_3$)$_2$CHCH$_2$OH

   **d.** $8.7 \times 10^{17}$ formula units Hg(C$_2$H$_3$O$_2$)$_2$

   **e.** $5.5 \times 10^{19}$ formula units Li$_2$CO$_3$

**11. a.** 52.9 g F$_2$

   **b.** $1.19 \times 10^3$ g or 1.19 kg BeSO$_4$

   **c.** $1.388 \times 10^5$ g or 138.8 kg CHCl$_3$

   **d.** $9.6 \times 10^{-12}$ g Cr(CHO$_2$)$_3$

   **e.** $6.6 \times 10^{-4}$ g HNO$_3$

   **f.** $2.38 \times 10^4$ g or 23.8 kg C$_2$Cl$_2$F$_4$

**12.** 0.158 mol Au

   0.159 mol Pt

   0.288 mol Ag

**13.** 0.234 mol C$_6$H$_5$OH

**14.** 3.8 g I$_2$

**15.** $1.00 \times 10^{22}$ atoms C

**16. a.** 0.0721 mol CaCl$_2$

   55.49 mol H$_2$O

   **b.** 0.0721 mol Ca$^{2+}$

   0.144 mol Cl$^-$

**17. a.** 1.325 mol C$_{12}$H$_{22}$O$_{11}$

   **b.** 7.762 mol NaCl

**18.** 0.400 mol ions

**19.** 4.75 mol atoms

**20. a.** 249 g H$_2$O

   **b.** 13.8 mol H$_2$O

   **c.** 36.1 mL H$_2$O

   **d.** 36.0 g H$_2$O

**21.** The mass of a sugar molecule is much greater than the mass of a water molecule. Therefore, the mass of 1 mol of sugar molecules is much greater than the mass of 1 mol of water molecules.

**22.** 1.52 g Al

**23.** 0.14 mol $O_2$

**24. a.** 0.500 mol Ag
0.250 mol S

**b.** 0.157 mol $Ag_2S$
0.313 mol Ag
0.157 mol S

**c.** 33.8 g Ag
5.03 g S

# PERCENTAGE COMPOSITION

**1. a.** $HNO_3$
1.60% H
22.23% N
76.17% O

**b.** $NH_3$
82.22% N
17.78% H

**c.** $HgSO_4$
67.616% Hg
10.81% S
21.57% O

**d.** $SbF_5$
56.173% Sb
43.83% F

**2. a.** 7.99% Li
92.01% Br

**b.** 94.33% C
5.67% H

**c.** 35.00% N
5.05% H
59.96% O

**d.** 2.15% H
29.80% N
68.06% O

**e.** 87.059% Ag
12.94% S

**f.** 32.47% Fe
13.96% C
16.29% N
37.28% S

**g.** $LiC_2H_3O_2$
10.52% Li
36.40% C
4.59% H
48.49% O

**h.** $Ni(CHO_2)_2$
39.46% Ni
16.15% C
1.36% H
43.03% O

**3. a.** 46.65% N

**b.** 23.76% S

**c.** 89.491% Tl

**d.** 39.17% O

**e.** 79.95% Br in $CaBr_2$

**f.** 78.767% Sn in $SnO_2$

**4. a.** 1.47 g O

**b.** 26.5 metric tons Al

**c.** 262 g Ag

**d.** 0.487 g Au

**e.** 312 g Se

**f.** $3.1 \times 10^4$ g Cl

**5. a.** 40.55% $H_2O$

**b.** 43.86% $H_2O$

**c.** 20.70% $H_2O$

**d.** 28.90% $H_2O$

**6. a.** $Ni(C_2H_3O_2)_2 \cdot 4H_2O$
23.58% Ni

**b.** $Na_2CrO_4 \cdot 4H_2O$
22.22% Cr

**c.** $Ce(SO_4)_2 \cdot 4H_2O$
34.65% Ce

**7.** 43.1 kg Hg

**8.** malachite: $5.75 \times 10^2$ kg Cu
chalcopyrite: $3.46 \times 10^2$ kg Cu
malachite has a greater Cu content

**9. a.** 25.59% V

**b.** 39.71% Sn

**c.** 22.22% Cl

**10.** 319.6 g anhydrous $CuSO_4$

**11.** 1.57 g $AgNO_3$

**12.** 54.3 g Ag
8.08 g S

**13.** 23.1 g $MgSO_4 \cdot 7H_2O$

**14.** $3.27 \times 10^2$ g S

# EMPIRICAL FORMULAS

**1. a.** $BaCl_2$

**b.** $BiO_3H_3$ or $Bi(OH)_3$

**c.** $AlN_3O_9$ or $Al(NO_3)_3$

**d.** $ZnC_4H_6O_4$ or $Zn(CH_3COO)_2$

**e.** $NiN_2S_2H_8O_8$ or $Ni(NH_4)_2SO_4$

**f.** $C_2HBr_3O_2$ or $CBr_3COOH$

**2. a.** $CuF_2$

**b.** $Ba(CN)_2$

**c.** $MnSO_4$

**3. a.** $NiI_2$

**b.** $MgN_2O_6$ or $Mg(NO_3)_2$

**c.** $MgS_2O_3$, magnesium thiosulfate

**d.** $K_2SnO_3$, potassium stannate

**4. a.** $As_2S_3$

**b.** $Re_2O_7$

**c.** $N_2H_4O_3$ or $NH_4NO_3$

**d.** $Fe_2Cr_3O_{12}$ or $Fe_2(CrO_4)_3$

**e.** $C_5H_9N_3$

**f.** $C_6H_5F_2N$ or $C_6H_3F_2NH_2$

5. **a.** $C_6H_{12}S_3$
   **b.** $C_8H_{16}O_4$
   **c.** $C_4H_6O_4$
   **d.** $C_{12}H_{12}O_6$
6. **a.** $C_4H_4O_4$
   **b.** $C_4H_8O_2$
   **c.** $C_9H_{12}O_3$
7. $K_2S_2O_5$, potassium metabisulfite
8. $Pb_3O_4$
9. $Cr_2S_3O_{12}$ or $Cr_2(SO_4)_3$, chromium(III) sulfate
10. $C_9H_6O_4$
11. $C_5H_9N_3$, the empirical formula and the molecular formula are the same
12. The molecular formulas of the compounds are different multiples of the same empirical formula. (FYI: The first could be acetic acid, $C_2H_4O_2$, and the second could be glucose, $C_6H_{12}O_6$, or some other simple sugar.)

# Stoichiometry

## STOICHIOMETRY

1. 15.0 mol $(NH_4)_2SO_4$
2. **a.** 51 g Al
   **b.** 101 g Fe
   **c.** 1.83 mol $Fe_2O_3$
3. 0.303 g $H_2$
4. $H_2SO_4 + 2KOH \rightarrow K_2SO_4 + 2H_2O$; 1.11 g $H_2SO_4$
5. **a.** $H_3PO_4 + 2NH_3 \rightarrow (NH_4)_2HPO_4$
   **b.** 0.293 mol $(NH_4)_2HPO_4$
   **c.** 970 kg $NH_3$
6. **a.** 90.0 mol $ZnCO_3$; 60.0 mol $C_6H_8O_7$
   **b.** 13.5 kg $H_2O$; 33.0 kg $CO_2$
7. **a.** 60.9 g methyl butanoate
   **b.** 3261 g $H_2O$
8. **a.** 0.450 mol $N_2$
   **b.** 294 g $NH_4NO_3$
9. $Pb(NO_3)_2 + 2KI \rightarrow PbI_2 + 2KNO_3$; 0.751 mg $KNO_3$
10. 3.3 mol $PbSO_4$
11. $2LiOH + CO_2 \rightarrow H_2O + Li_2CO_3$; 360 g $H_2O$
12. **a.** 38.1 g $H_2O$
    **b.** 40.1 g $H_3PO_4$
    **c.** 0.392 mol $H_2O$
13. $C_2H_5OH + 3O_2 \rightarrow 2CO_2 + 3H_2O$; 81.0 g $C_2H_5OH$
14. 76.5 g $H_2SO_4$; 12.5 g $O_2$
15. $2NaHCO_3 \rightarrow Na_2CO_3 + H_2O + CO_2$; 1.31 g $CO_2$

16. **a.** $2N_2H_4 + N_2O_4 \rightarrow 3N_2 + 4H_2O$
    **b.** 1 mol $N_2O_4$ to 3 mol $N_2$
    **c.** 30 000 mol $N_2$
    **d.** $3.52 \times 10^5$ g $H_2O$
17. $2HgO(s) \rightarrow 2Hg(l) + O_2(g)$; 1.1954 mol $O_2$
18. $2Fe + 3Cl_2 \rightarrow 2FeCl_3$; 30.5 g Fe
19. 9.26 mg CdS
20. **a.** 1.59 mol $CO_2$
    **b.** 0.0723 mol $C_3H_5(OH)_3$
    **c.** 535 g $Mn_2O_3$
    **d.** 8.33 g $C_3H_5(OH)_3$; 4.97 g $CO_2$
21. **a.** $3.29 \times 10^3$ kg of HCl
    **b.** 330 g $CO_2$ (s)
22. **a.** $6.53 \times 10^5$ g $NH_4ClO_4$
    **b.** 160 kg $NO(g)$
23. **a.** $1.70 \times 10^6$ mol $H_3PO_4$
    **b.** 666 kg of $CaSO_4 \cdot 2H_2O$
    **c.** 34 metric tons of $H_3PO_4$
24. 1670 kg

## LIMITING REACTANTS

1. $2ZnS + 3O_2 \rightarrow 2ZnO + 2SO_2$; ZnS is limiting
2. **a.** Al is limiting
   **b.** $4.25 \times 10^{-3}$ mol $Al_2O_3$
   **c.** $O_2$ is limiting
3. **a.** CuS is limiting
   **b.** 15.6 g CuO
4. Fe is limiting; 0.158 mol Cu
5. 54 g $Ba(NO_3)_2$
6. **a.** 38 g $Br_2$
   **b.** 510 g $I_2$
7. **a.** Ni is in excess
   **b.** 60.2 g $Ni(NO_3)_2$
8. $CS_2(g) + 3O_2(g) \rightarrow 2SO_4(g) + CO_2(g)$ 0.80 mol $O_2$ remain
9. **a.** 0.84 g $Hg(NH_2)Cl$
   **b.** 0.84 g
10. **a.** $2Al(s) + 2NaOH(aq) + 2H_2O(l) \rightarrow 2NaAlO_2(aq) + 3H_2(g)$
    **b.** NaOH is limiting; 0.56 mol $H_2$
    **c.** Al should be limiting because you would not want aluminum metal remaining in the drain.
11. **a.** 0.0422 mol Cu; 0.169 mol $HNO_3$
    **b.** Cu is in excess
    **c.** 3.32 g $H_2O$
12. **a.** 2.90 mol NO; 4.35 mol $H_2O$
    **b.** $NH_3$ is limiting
    **c.** $NH_3$ is limiting; $1.53 \times 10^3$ kg NO

**13.** 565 g $CH_3CHO$;
29 g $CH_3CH_2OH$ remains
**14.** 630 g HBr
**15.** 12.7 g $SO_2$
**16. a.** 18.4 g Tb
   **b.** 2.4 g $TbF_3$

## PERCENTAGE YIELD

**1. a.** 64.3% yield
   **b.** 58.0% yield
   **c.** 69.5% yield
   **d.** $CH_3CH_2OH$ is limiting; 79% yield
**2. a.** 69.5% yield
   **b.** 79.0% yield
   **c.** 48% yield
   **d.** 85% yield
**3. a.** 59% yield
   **b.** 81.0% yield
   **c.** $2.3 \times 10^5$ mol P
**4. a.** 91.8% yield
   **b.** 0.0148 mol W
   **c.** 16.1 g $WO_3$
**5. a.** 86.8% yield
   **b.** 92.2% yield
   **c.** $2.97 \times 10^4$ kg $CS_2$;
     $4.39 \times 10^4$ kg $S_2Cl_2$
**6. a.** 81% yield
   **b.** $2.0 \times 10^2$ g $N_2O_5$
**7.** 80.1% yield
**8. a.** 95% yield
   **b.** $9.10 \times 10^2$ g Au
   **c.** $9 \times 10^4$ kg ore
**9. a.** 87.5% yield
   **b.** 0.25 g CO
**10. a.** 71% yield
   **b.** 26 metric tons
   **c.** 47.8 g NaCl
   **d.** 500 kg per hour NaOH
**11. a.** $2Mg + O_2 \rightarrow 2MgO$
   **b.** 87.7% yield
   **c.** $3Mg + N_2 \rightarrow Mg_3N_2$
   **d.** 56% yield
**12. a.** 80.% yield
   **b.** 66.2% yield
   **c.** 57.1% yield
**13.** $2C_3H_6(g) + 2NH_3(g) + 3O_2(g) \rightarrow$
$2C_3H_3N(g) + 6H_2O(g)$
91.0% yield
**14. a.** $CO + 2H_2 \rightarrow CH_3OH$
     $3.41 \times 10^3$ kg
   **b.** 91.5% yield
**15.** 96.9% yield

**16.** $6CO_2 + 6H_2O \rightarrow C_6H_{12}O_6 + 6O_2$
$6.32 \times 10^3$ g $O_2$
**17.** 27.6 kg CaO

# Causes of Change

## THERMOCHEMISTRY

**1.** $-260.8$ kJ/mol
**2.** $-385.9$ kJ/mol
**3.** $-390.$ kJ/mol
**4.** $-492.3$ kJ/mol
**5.** $-107.6$ kJ/mol
**6.** $-121.8$ kJ/mol
**7.** $-384.9$ kJ/mol
**8.** 74.2 kJ/mol
**9.** $-169.0$ kJ/mol
**10. a.** $CH_4(g) + 2O_2(g) \rightarrow$
     $CO_2(g) + 2H_2O(g)$
     $C_3H_8(g) + 5O_2(g) \rightarrow$
     $3CO_2(g) + 4H_2O(g)$
   **b.** for methane: $\Delta H = -802.2$ kJ/mol
     for propane: $\Delta H = -2043$ kJ/mol
   **c.** $output_{methane} = -4.998 \times 10^4$ kJ/kg
     $output_{propane} = -4.632 \times 10^4$ kJ/kg
     Methane yields more energy per mass.
**11.** $-132.7$ kJ/mol
**12.** $-7171.4$ kJ/mol
**13.** $-141.1$ kJ/mol
**14.** 20.2 kJ/mol
**15.** $-285$ kJ/mol·K
**16. a.** 786.8 kJ/mol
   **b.** $-36$ kJ/mol
   **c.** 2154 kJ/mol
   **d.** $-496$ kJ/mol
   **e.** 1346.4 kJ/mol

# Gases

## GAS LAWS

**1. a.** 105 kPa
   **b.** 5.0 mL
   **c.** 42.4 kPa
   **d.** $6.78 \times 10^{-3}$ dm$^3$
   **e.** 1.24 atm
   **f.** 1.5 m$^3$
**2.** 8.0 m$^3$
**3.** 0.0258 atm
**4.** $8.01 \times 10^{-2}$ dm$^3$
**5. a.** 234 K
   **b.** 1.2 dm$^3$
   **c.** $-269.17°C$
   **d.** $8.10 \times 10^{-2}$ L
   **e.** 487 cm$^3$
   **f.** 67.9 m$^3$

6. 1.45 cm$^3$
7. −40.°C
8. a. −208.6°C
   b. 5.5 kPa
   c. 2.61 atm
   d. 297°C
   e. 35.6 atm
   f. 39 K
9. 0.899 atm
10. 2.23 atm
11. 7.98 K
12. a. 2.02 L
    b. 75.8 kPa
    c. 110 K
    d. 4.69 × 10$^3$ mm$^3$
    e. −72°C
    f. 2.25 atm
13. 379 cm$^3$
14. 98 kPa
15. 1.00 atm; use Boyle's law to find the pressure of each gas in the whole space; add the partial pressures of both gases when they occupy the whole space.
16. 285 mL
17. 89 cm$^3$ The pressure in the bottle on top of the mountain is the sum of $P_{O_2\ dry}$ at the temperature of the mountaintop and $P_{H_2O\ vapor}$ at the temperature on top of the mountain.
18. 59 cm$^3$
    Solve the equation:
    $V_{total}/293\ K = (V_{total} + 0.20\ cm^3)/294\ K$
19. 118 kPa
20. 935 mL
21. 26.4 kL
22. 3.76 atm
23. 115 mL
24. 10.9 atm

## IDEAL GAS LAW

1. a. 34.2 mol
   b. 6.68 × 10$^3$ kPa
   c. −148°C
   d. 1.1 × 10$^5$ L
2. a. 55.9 g/mol
   b. 0.111 g
   c. 4.46 × 10$^{-2}$ L
   d. 0.846 atm
   e. 391 g/mol
3. 2.71 × 10$^3$ L
4. a. 48.4 g/mol
   b. 9.38 g/L

c. 6°C
d. 2.24 atm or 227 kPa
5. 33.7 atm
6. 3.06 g/L
7. 663 g/mol
8. 204 L
9. 0.0101 mol ethane
10. 5.16 g NO
11. 77.0% yield
12. 10.5 g/mol
13. 171 g/mol
14. 6.55 atm
15. 326 kPa
16. 479 K or 206°C
17. 1210 L at −75°C; 1620 L at −8°C
18. 1.11 × 10$^4$ kPa
19. 168 mL
20. 3.85 × 10$^3$ L
21. 4.05 × 10$^3$ L
22. 29.0 g/mol

## STOICHIOMETRY OF GASES

1. a. 19.0 mL N$_2$
   b. 4.26 × 10$^4$ L NH$_3$
   c. 896 L NH$_3$
   d. 899 L H$_2$
2. a. 23 L H$_2$O
   b. 1070 L O$_2$
   c. 326 mL CO$_2$
   d. 25.2 L total products
3. 1550 L O$_2$ at STP
4. 0.894 L SiF$_4$
5. a. 3.36 L H$_2$
   b. 488 g Fe
   c. 112 L H$_2$
6. 0.013 L H$_2$ or 13 mL H$_2$
7. 7.50 L O$_2$ at STP; 4.14 g diethyl ether
8. a. 3.36 L N$_2$; 6.72 L CO$_2$; 5.60 L H$_2$O; 0.560 L O$_2$
   b. 15.0 L total volume all gases
9. 0.894 g NH$_4$NO$_3$
10. 2.19 L PH$_3$
11. 1.2 × 10$^3$ kg Al; 7.4 × 10$^5$ L HCl
12. 7.08 × 10$^7$ L NH$_3$
13. 3.77 g BaO$_2$
14. 5.85 L Cl$_2$
15. 28.0 kL NH$_3$; 28.0 kL NO$_2$ overall reaction is: $4NH_3 + 7O_2 \rightarrow 4NO_2 + 6H_2O$
16. 18.2 g KClO$_3$
17. a. 38 000 L NH$_3$
    b. 1.30 × 10$^5$ g NaHCO$_3$
    c. 12.3 L NH$_3$
    d. 5.60 × 10$^3$ L

**18. a.** 9.93 L $CO_2$
  **b.** 1.63 L $O_2$
  **c.** 26.0 L $CO_2$
  **d.** $3.29 \times 10^3$ g $H_2O$

# Solutions

## CONCENTRATION OF SOLUTIONS

**1. a.** 60.0 g $KMnO_4$; 440.0 g $H_2O$
  **b.** 220 g $BaCl_2$
  **c.** 457 g glycerol
  **d.** 0.0642 M $K_2Cr_2O_7$
  **e.** 1.27 m $CaCl_2$
  **f.** 0.234 g NaCl
  **g.** 541 g glucose; 2040 g total
**2.** 10.6 mol $H_2SO_4$
**3.** 0.486 m linoleic acid
**4. a.** 13.0 g $Na_2S_2O_3$
  **b.** 0.0820 mol
  **c.** 0. 328 M
**5.** 338 g $CoCl_2$
**6.** 0.442 L
**7.** 203 g urea
**8.** 18.8 g $Ba(NO_3)_2$
**9.** add 3.5 g $(NH_4)_2SO_4$ to 96.5 g $H_2O$
**10.** 54 g $CaCl_2$
**11.** 1.25 mol; 1.25 M
**12.** 93.6 g/mol
**13.** 49.6 kg water; 0.5 kg NaCl
**14.** 8.06%
**15.** 1.4 L ethyl acetate
**16.** $CdCl_2(aq) + Na_2S(aq) \rightarrow CdS(s) + 2$ $NaCl(aq)$
  **a.** 0.196 mol $CdCl_2$
  **b.** 0.196 mol CdS
  **c.** 28.3 g CdS
**17.** 34.4 g $H_2SO_4$
**18.** $1.54 \times 10^5$ mol HCl
**19.** 85.7 mL $BaCl_2$ solution
**20. a.** Measure out 9.39 g $CuSO_4 \cdot 5H_2O$ and add 90.61 g $H_2O$ to make 100. g of solution. The 9.39 g of $CuSO_4 \cdot 5H_2O$ contributes the 6.00 g of $CuSO_4$ needed.
  **b.** Measure out 200. g $CuSO_4 \cdot 5H_2O$, dissolve in water, then add water to make 1.00 L. Water of hydration does not have to be considered here as long as the molar mass of the hydrate is used in determining the mass to weigh out.
  **c.** Measure out 870 g $CuSO_4 \cdot 5H_2O$ and add 685 g $H_2O$. The hydrate

contributes 315 g of $H_2O$, so only 685 g $H_2O$ must be added.
**21.** 383 g $CaCl_2 \cdot 6H_2O$
**22.** 0.446 g arginine
**23.** 3254 g $H_2O$; the hydrate contributes 987.9 g $H_2O$.
**24.** 9.646 g $KAl(SO_4)_2 \cdot 12H_2O$; 25.35 g $H_2O$

## DILUTIONS

**1.** 0.0948 M
**2.** 0.44 mL
**3. a.** 3.0 M
  **b.** 0.83 L
  **c.** $1.5 \times 10^3$ g
**4.** 6.35 mL
**5.** 348 mL
**6.** 0.558 M
**7. a.** 850 mL; 2.4 mL; 86 mL
  **b.** 1.3 L concentrated $HNO_3$
  **c.** 1.16 L concentrated HCl
**8.** 0.48 M
**9.** 2.72 M
**10.** Dilute 4.73 mL of the 6.45 M acetic acid to 25 mL. This means adding 20.27 mL of water.
**11.** 39.7 g/mol
**12.** 0.27 M
**13.** $1.9 \times 10^2$ g
**14.** 0.667 L
**15.** For 1.00 L of 0.495 M urea solution, take 161 mL of 3.07 M stock solution of urea and dilute with water (839 mL of water) to make 1.00 L.
**16. a.** 17.8 m
  **b.** 1080 g, 6.01 M
**17.** 47.17 g $Na_2CO_3$ per 50.00 g sample = 94.3% $Na_2CO_3$
**18.** 151 g $CuCl_2$
**19.** Add 2.3 volumes of $H_2O$ per volume of stock solution.
**20. a.** 14.9 g
  **b.** $7.02 \times 10^{-2}$ mol
  **c.** 0.167 M

## COLLIGATIVE PROPERTIES

**1.** $-9.88°C$ ; $102.7°C$
**2.** $103.3°C$
**3.** $201.6°C$
**4.** 50. g ethanol
**5.** 82 g/mol
**6.** $-18.2°C$
**7.** 15.5 g/mol
**8.** 66.8 g/mol
**9.** $183.3°C$

10. **a.** $-15.0°C$
    **b.** $104.1°C$
11. $292$ g/mol
12. $-47.0°C$
13. $190$ g/mol
14. $107.2°C$
15. $-27.9°C$
16. $204.4$ g/mol; $C_{16}H_{10}$
17. $9.9$ g $CaCl_2$; $49$ g glucose
18. $C_3H_6O_3$
19. $-29.2°C$
20. $2.71$ kg; $104.9°C$
21. **a.** $2.2$ m
    **b.** $1.7$ m
22. $92.0$ g $H_2O$
23. $150$ g/mol; $C_5H_{10}O_5$
24. $124$ g/mol; $C_6H_6O_3$

# Chemical Equilibrium

## EQUILIBRIUM

1. $1.98 \times 10^{-7}$
2. $2.446 \times 10^{-12}$
3. $3.97 \times 10^{-5}$
4. **a.** $8.13 \times 10^{-4}$ M
   **b.** $0.0126$ M
5. **a.** The concentrations are equal.
   **b.** $K$ will increase.
6. $0.09198$ M
7. **a.** $0.7304$ M
   **b.** $6.479 \times 10^{-4}$ M
8. **a.** $[A] = [B] = [C] = 1/2[A]_{initial}$
   **b.** $[A]$, $[B]$, and $[C]$ will increase equally. K remains the same.
9. **a.** $K_{eq} = [HBr]^2/[H_2][Br_2]$
   **b.** $2.11 \times 10^{-10}$ M
   **c.** $Br_2$ and $H_2$ will still have the same concentration. HBr will have a much higher concentration than the two reactants; at equilibrium, essentially only HBr will be present.
10. $1.281 \times 10^{-6}$
11. $4.61 \times 10^{-3}$
12. **a.** $K_{eq} = [HCN]/[HCl]$
    **b.** $3.725 \times 10^{-7}$ M
13. **a.** The reaction yields essentially no products at 25°C; as a result, the equilibrium constant is very small. At 110 K, the reaction proceeds to some extent.
    **b.** $2.51$ M
14. $0.0424$
15. $0.0390$

# EQUILIBRIUM OF SALTS, $K_{sp}$

1. $1.0 \times 10^{-4}$
2. $1.51 \times 10^{-7}$
3. **a.** $1.6 \times 10^{-11}$
   **b.** $0.49$ g
4. $2.000 \times 10^{-7}$
5. **a.** $8.9 \times 10^{-4}$ M
   **b.** $0.097$ g
6. $0.036$ M
7. **a.** $1.1 \times 10^{-4}$ M
   **b.** $4.6$ L
8. **a.** $5.7 \times 10^{-4}$ M
   **b.** $2.1$ g
9. $0.83$ g remains
10. $2.8 \times 10^{-3}$ M
11. **a.** $1.2 \times 10^{-5}$
    **b.** Yes
12. **a.** $1.1 \times 10^{-8}$
    **b.** No
13. **a.** $Mg(OH)_2 \rightarrow Mg^{2+} + 2OH^-$
    **b.** $11$ L
    **c.** A suspension contains undissolved $Mg(OH)_2$ suspended in a saturated solution of $Mg(OH)_2$. As hydroxide ions are depleted by titration, the dissociation equilibrium continues to replenish them until all of the $Mg(OH)_2$ is used up.
14. **a.** $0.184$ M
    **b.** $46.8$ g
15. $5.040 \times 10^{-3}$
16. **a.** $1.0$
    **b.** $0.25$
    **c.** $0.01$
    **d.** $1 \times 10^{-6}$
17. $4 \times 10^{-4}$; $3 \times 10^{-5}$; $3 \times 10^{-5}$; $1 \times 10^{-6}$; $3 \times 10^{-7}$
18. $7.1 \times 10^{-7}$ M; $1.3 \times 10^{-3}$ g
19. **a.** $0.011$ M
    **b.** $0.022$ M
    **c.** $12.35$
20. $0.063$ g

# Acids and Bases

## pH

1. **a.** $[OH^-] = 0.05$ M, $[H_3O^+] = 2 \times 10^{-13}$ M
   **b.** $[OH^-] = 2.0 \times 10^{-12}$ M, $[H_3O^+] = 5.0 \times 10^{-3}$ M
   **c.** $[OH^-] = 0.013$ M, $[H_3O^+] = 7.7 \times 10^{-13}$ M

**d.** $[OH^-] = 6.67 \times 10^{-14}$ M,
   $[H_3O^+] = 0.150$ M

**e.** $[OH^-] = 0.0400$ M,
   $[H_3O^+] = 2.50 \times 10^{-13}$ M

**f.** $[OH^-] = 2.56 \times 10^{-14}$ M,
   $[H_3O^+] = 0.390$ M

**g.** 10, 2.3, 12.11, 0.824, 12.602, 0.409

**2.** $[OH^-] = 0.160$ M
   $[H_3O^+] = 6.25 \times 10^{-14}$ M

**3.** 0.08 M

**4.** 2.903; 4.137

**5.** 10.0; 10.7

**6. a.** $[H_3O^+] = 0.020$ M,
   $[OH^-] = 5.0 \times 10^{-13}$ M

   **b.** 1.7

**7. a.** 2.804

   **b.** 2.755

**8.** $2.9 \times 10^{-11}$ M

**9.** 10.54

**10.** $[H_3O^+] = 1 \times 10^{-4}$ M
   $[OH^-] = 1 \times 10^{-10}$ M

**11.** 10.96

**12.** 0.2 g

**13.** $[H_3O^+] = 1.0 \times 10^{-3}$ M
   $[OH^-] = 1.0 \times 10^{-4}$ M

**14. a.** 0.40%

   **b.** 1.3%

   **c.** 4.0%

**15.** 1.402

**16. a.** 4.50

   **b.** 2.40

   **c.** 3.83

   **d.** 5.63

**17.** $[H_3O^+] = 1 \times 10^{-9}$ M
   $[OH^-] = 1 \times 10^{-5}$ M

**18.** 11.70

**19.** $[H_3O^+] = 0.020$ M
   $[HCl] = 0.020$ M

**20.** $3.2 \times 10^{-4}$ M

**21.** pH = 0.000
   10. kL
   $1.0 \times 10^2$ L
   10. kL

**22.** 0.052 mol NaOH
   2.1 g NaOH

**23.** $7.1 \times 10^{-3}$

**24.** $[H_3O^+] = 7.1 \times 10^{-4}$ M
   $[OH^-] = 1.4 \times 10^{-11}$ M

## TITRATIONS

**1.** 0.4563 M KOH

**2.** 2.262 M $CH_3COOH$

**3.** 2.433 M $NH_3$

**4. a.** 20.00 mL base

   **b.** 10.00 mL acid

   **c.** 80.00 mL acid

**5.** 5.066 M HF

**6.** 0.2216 M oxalic acid

**7.** 1.022 M $H_2SO_4$

**8.** 0.1705 M KOH

**9.** 0.5748 M citric acid

**10.** 0.3437 M KOH

**11.** 43.2 mL NaOH

**12.** 23.56 mL $H_2SO_4$

**13.** 0.06996 mol KOH; 3.926 g KOH;
   98.02% KOH

**14.** 51.7 g $Mg(OH)_2$

**15.** 0.514 M $NH_3$

**16.** 20.85 mL oxalic acid

**17. a.** 0.5056 M HCl

   **b.** 0.9118 M RbOH

**18.** 570 kg $Ca(OH)_2$

**19.** 16.9 M $HNO_3$

**20.** 33.58 mL

## EQUILIBRIUM OF ACIDS AND BASES, $K_a$ AND $K_b$

**1.** pH = 2.857
   $K_a = 1.92 \times 10^{-5}$

**2.** pH = 1.717
   $K_a = 2.04 \times 10^{-3}$

**3. a.** $[H_3O^+] = 3.70 \times 10^{-11}$ M
   pH = 10.431
   $K_b = 1.82 \times 10^{-7}$

   **b.** $[B] = 4.66 \times 10^{-3}$ M
   $K_b = 1.53 \times 10^{-4}$
   pH = 10.93

   **c.** $[OH^-] = 1.9 \times 10^{-3}$ M
   $[H_3O^+] = 5.3 \times 10^{-12}$ M
   $[B] = 0.0331$ M
   $K_b = 1.1 \times 10^{-4}$

   **d.** $[B]_{initial} = 7.20 \times 10^{-3}$ M
   $K_b = 1.35 \times 10^{-4}$
   pH = 10.96

**4.** $6.4 \times 10^{-5}$

**5.** $K_b = 3.1 \times 10^{-5}$
   $[H_2NCH_2CH_2OH] = 6.06 \times 10^{-3}$ M

**6.** $1.3 \times 10^{-4}$

**7.** $[OH^-] = 1.58 \times 10^{-5}$ M
   pH = 9.20

**8.** $K_a = 2.2 \times 10^{-3}$
   $[HB]_{initial} = 0.0276$ M

**9.** $4.63 \times 10^{-3}$

**10.** $2.62 \times 10^{-4}$

**11.** $[H_3O^+] = 0.0124$ M
   pH = 1.907

**12.** $[H_3O^+] = 0.0888$ M

pH = 1.05

**13.** $K_a = 3.50 \times 10^{-4}$

pH = 2.241

**14. a.** $3.16 \times 10^{-6}$ M

**b.** 5.500

**c.** 0.025 M

**d.** $4.0 \times 10^{-10}$

**e.** HCN is a fairly weak acid.

**f.** $5.8 \times 10^{+6}$ M

**15. a.** 0.740

**b.** 0.0325

**c.** 1.02 M

**d.** It is moderately weak.

**16.** $2.80 \times 10^{-11}$

**17. a.** 0.200 M

**b.** 0.233

**18.** pH = 11.37

$K_b = 7.41 \times 10^{-5}$

**19.** $7.36 \times 10^{-4}$

dimethylamine is the stronger base

**20. a.** $8.3 \times 10^{-7}$

**b.** $1.3 \times 10^{-4}$

**c.** 10.11

# Oxidation, Reduction, and Electronegativity

## REDOX EQUATIONS

**1.** $3Mg + N_2 \rightarrow Mg_3N_2$

**2.** $SO_2 + Br_2 + 2H_2O \rightarrow 2HBr + H_2SO_4$

**3.** $H_2S + Cl_2 \rightarrow S + 2HCl$

**4.** $PbO_2 + 4HBr \rightarrow PbBr_2 + Br_2 + 2H_2O$

**5.** $S + 6HNO_3 \rightarrow 6NO_2 + H_2SO_4 + 2H_2O$

**6.** $NaIO_3 + N_2H_4 + 2HCl \rightarrow$
$N_2 + NaICl_2 + 3H_2O$

**7.** $MnO_2 + H_2O_2 + 2HCl \rightarrow$
$MnCl_2 + O_2 + 2H_2O$

**8.** $3AsH_3 + 4NaClO_3 \rightarrow$
$3H_3AsO_4 + 4NaCl$

**9.** $K_2Cr_2O_7 + 3H_2C_2O_4 + 8HCl \rightarrow$
$2CrCl_3 + 6CO_2 + 2KCl + 7H_2O$

**10.** $2Hg(NO_3)_2 \rightarrow 2HgO + 4NO_2 + O_2$

**11.** $4HAuCl_4 + 3N_2H_4 \rightarrow$
$4Au + 3N_2 + 16HCl$

**12.** $5Sb_2(SO_4)_3 + 4KMnO_4 + 24H_2O \rightarrow$
$10H_3SbO_4 + 2K_2SO_4 + 4MnSO_4 +$
$9H_2SO_4$

**13.** $3Mn(NO_3)_2 + 5NaBiO_3 + 9HNO_3 \rightarrow$
$5Bi(NO_3)_2 + 3HMnO_4 + 5NaNO_3 +$
$3H_2O$

**14.** $H_3AsO_4 + 4Zn + 8HCl \rightarrow$
$AsH_3 + 4ZnCl_2 + 4H_2O$

**15.** $KClO_3 + 6HCl \rightarrow 3Cl_2 + 3H_2O + KCl$

**16.** $2KClO_3 + 4HCl \rightarrow$
$Cl_2 + 2ClO_2 + 2H_2O + 2KCl$

**17.** $2MnCl_3 + 2H_2O \rightarrow$
$MnCl_2 + MnO_2 + 4HCl$

**18.** $2NaOH + 6H_2O + 2Al \rightarrow$
$2NaAl(OH)_4 + 3H_2$

**19.** $6Br_2 + 6Ca(OH)_2 \rightarrow$
$5CaBr_2 + Ca(BrO_3)_2 + 6H_2O$

**20.** $N_2O + 2NaClO + 2NaOH \rightarrow$
$2NaCl + 2NaNO_2 + H_2O$

**21.** $4HBr + MnO_2 \rightarrow$
$MnBr_2 + 2H_2O + Br_2$

**22.** $Au + 4HCl + HNO_3 \rightarrow$
$HAuCl_4 + NO + 2H_2O$

## ELECTROCHEMISTRY

**1.** $E^0 = +2.77$ V; spontaneous

**2.** $E^0 = +1.11$ V; spontaneous

**3.** $E^0 = -0.46$ V; not spontaneous

**4.** $E^0 = +1.50$ V; spontaneous

**5.** $E^0 = +2.46$ V; spontaneous

**6.** $E^0 = +1.28$ V; spontaneous

**7.** $E^0 = +3.71$ V; spontaneous

**8.** $E^0 = -3.41$ V; not spontaneous

**9.** $E^0 = +1.32$ V; spontaneous

**10.** $E^0 = -3.60$ V; not spontaneous

**11.** Overall reaction:
$Cl_2 + Ni \rightarrow Ni^{2+} + 2Cl^-$
Cathode reaction: $Cl_2 + 2e^- \rightarrow 2Cl^-$
Anode reaction: $Ni \rightarrow Ni^{2+} + 2e^-$
Cell voltage: $+1.62$ V

**12.** Overall reaction:
$3Hg^{2+} + 2Fe \rightarrow 3Hg + 2Fe^{3+}$
Cathode reaction: $Hg^{2+} + 2e^- \rightarrow Hg$
Anode reaction: $Fe \rightarrow Fe^{3+} + 3e^-$
Cell voltage: $+0.89$ V

**13.** Overall reaction:
$3MnO_4^- + Al \rightarrow 3MnO_4^{2-} + Al^{3+}$
Cathode reaction:
$MnO_4^- + e^- \rightarrow MnO_4^{2-}$
Anode reaction: $Al \rightarrow Al^{3+} + 3e^-$
Cell voltage: $+2.22$ V

**14.** Overall reaction:
$2MnO_4^- + 6H^+ + 5H_2S \rightarrow$
$2Mn^{2+} + 8H_2O + 5S$
Cathode reaction:
$MnO_4^- + 8H^+ + 5e^- \rightarrow Mn^{2+} + 4H_2O$
Anode reaction: $H_2S \rightarrow S + 2H^+ + 2e^-$
Cell voltage: $+1.36$ V

**15.** Overall reaction:
$Ca^{2+} + 2Li \rightarrow Ca + 2Li^+$
Cathode reaction: $Ca^{2+} + 2e^- \rightarrow Ca$

Anode reaction: $Li \rightarrow Li^+ + e^-$
Cell voltage: $+0.17$ V

16. Overall reaction:
$2MnO_4^- + 16H^+ + 10Br^- \rightarrow$
$2Mn^{2+} + 8H_2O + 5Br_2$
Cathode reaction:
$MnO_4^- + 8H^+ + 5e^- \rightarrow$
$Mn^{2+} + 4H_2O$
Anode reaction: $2Br^- \rightarrow Br_2 + 2e^-$
Cell voltage: $+0.43$ V

17. Overall reaction:
$2Fe^{3+} + Sn \rightarrow 2Fe^{2+} + Sn^{2+}$
Cathode reaction: $Fe^{3+} + e^- \rightarrow Fe^{2+}$
Anode reaction: $Sn \rightarrow Sn^{2+} + 2e^-$
Cell voltage: $+0.91$ V

18. Overall reaction:
$Cr_2O_7^{2-} + 14H^+ + 3Zn \rightarrow$
$2Cr^{3+} + 7H_2O + 3Zn^{2+}$
Cathode reaction:
$Cr_2O_7^{2-} + 14H^+ + 6e^- \rightarrow$
$2Cr^{3+} + 7H_2O$
Anode reaction: $Zn \rightarrow Zn^{2+} + 2e^-$
Cell voltage: $+1.99$ V

19. Overall reaction:
$Ba + Ca^{2+} \rightarrow Ba^{2+} + Ca$
Cathode reaction: $Ca^{2+} + 2e^- \rightarrow Ca$
Anode reaction: $Ba \rightarrow Ba^{2+} + 2e^-$
Cell voltage: $+0.04$ V

20. Overall reaction:
$Cd + Hg_2^{2+} \rightarrow Cd^{2+} + 2Hg$
Cathode reaction: $Hg_2^{2+} + 2e^- \rightarrow 2Hg$
Anode reaction: $Cd \rightarrow Cd^{2+} + 2e^-$
Cell voltage: $+1.20$ V